Trends in Mathematics is a series devoted to the publication of volumes arising from conferences and lecture series focusing on a particular topic from any area of mathematics. Its aim is to make current developments available to the community as rapidly as possible without compromise to quality and to archive these for reference.

Proposals for volumes can be sent to the Mathematics Editor at either

Birkhäuser Verlag
P.O. Box 133
CH-4010 Basel
Switzerland

or

Birkhäuser Boston Inc.
675 Massachusetts Avenue
Cambridge, MA 02139
USA

Material submitted for publication must be screened and prepared as follows:

All contributions should undergo a reviewing process similar to that carried out by journals and be checked for correct use of language which, as a rule, is English. Articles without proofs, or which do not contain any significantly new results, should be rejected. High quality survey papers, however, are welcome.

We expect the organizers to deliver manuscripts in a form that is essentially ready for direct reproduction. Any version of T_EX is acceptable, but the entire collection of files must be in one particular dialect of T_EX and unified according to simple instructions available from Birkhäuser.

Furthermore, in order to guarantee the timely appearance of the proceedings it is essential that the final version of the entire material be submitted no later than one year after the conference. The total number of pages should not exceed 350. The first-mentioned author of each article will receive 25 free offprints. To the participants of the congress the book will be offered at a special rate.

Bifurcation, Symmetry and Patterns

Jorge Buescu
Sofia Castro
Ana Paula Dias
Isabel Labouriau
Editors

Springer Basel AG

Editors' addresses:

Jorge Buescu
Centro de Análise Matemática
Geometria e Sistemas Dinâmicos
Departamento de Matemática
Instituto Superior Técnico
Av. Rovisco Pais
1049-001 Lisboa, Portugal
jbuescu@math.ist.utl.pt

Sofia B.S.D. Castro
Centro de Matemática Aplicada
Faculdade de Economia
Universidade do Porto
Rua Dr. Roberto Frias
4200 Porto, Portugal
sdcastro@fep.up.pt

Ana Paula da Silva Dias
Centro de Matemática
Departamento de Matemática Pura
Universidade do Porto
Praça Gomes Teixeira
Faculdade de Ciências
Universidade do Porto
4099-002 Porto, Portugal
apdias@fc.up.pt

Isabel Salgado Labouriau
Centro de Matemática Aplicada
Departamento de Matemática Aplicada
Rua das Taipas, 135
4050-600 Porto, Portugal
islabour@fc.up.pt

2000 Mathematical Subject Classification 34-06, 35-06, 37-06, 92-06

A CIP catalogue record for this book is available from the
Library of Congress, Washington D.C., USA

Bibliographic information published by Die Deutsche Bibliothek
Die Deutsche Bibliothek lists this publication in the Deutsche Nationalbibliografie; detailed
bibliographic data is available in the Internet at <http://dnb.ddb.de>.

ISBN 978-3-0348-9642-9   ISBN 978-3-0348-7982-8 (eBook)

DOI 10.1007/978-3-0348-7982-8

© 2003 Springer Basel AG
Originally published by Birkhäuser Verlag, Basel – Boston – Berlin in 2003
Softcover reprint of the hardcover 1st edition 2003

Front cover picture: The image "Porto" is a graphical representation of a symmetric chaotic
attractor with five-fold rotational symmetry. It was created specially for the conference by
Mike Field (University of Houston).
Printed on acid-free paper produced from chlorine-free pulp. TCF ∞

ISBN 978-3-0348-9642-9

9 8 7 6 5 4 3 2 1   www.birkhauser-science.com

Contents

vi   Contents

Foreword by the Editors

Martin Golubitsky and Ian Stewart started their mathematical lives in 1970 and 1969 by looking at Lie groups and Lie algebras, respectively. For a decade, their work contributes to differential geometry and algebra, focussing on Lie theory, symplectic geometry and singularity theory.

In 1978, Marty publishes a paper with the title "An introduction to catastrophe theory and its applications" and in 1980, the same words "catastrophe theory" appear in the title of a paper by Ian, "Catastrophe theory and equations of state: conditions for a butterfly singularity". They had met not long before. From this point onwards, both Marty and Ian become concerned with bifurcation theory and their first joint publication appears in 1984 in the Bulletin of the AMS, "Hopf bifurcation in the presence of symmetry". Their publications total more than 200 papers as well as the commonly referred to as "volume 1" and "volume 2", whose full titles are "Singularities and Groups in Bifurcation Theory – vol. 1" by Golubitsky and Schaeffer and "Singularities and Groups in Bifurcation Theory – vol. 2" by Golubitsky, Stewart and Schaeffer, both published by Springer. Volume 2 is said to be the book that inspired the largest number of PhD theses ever. What may or may not be referred to as "volume 3" has now been published by Birkhäuser under the title "The Symmetry Perspective: From equilibrium to chaos in phase space and physical space". This volume was produced by Ian and Marty in part for a summer school, organized by Castro, Dias and Labouriau, which followed the conference in their honour whose Proceedings are hereby presented. The text used by Marty and Ian was then developed to win the 2001 Ferran Sunyer i Balaguer Prize.

Other than with their own research, Marty and Ian contributed to the dissemination of state of the art mathematics by the edition of several volumes of Conference Proceedings. They have organized or helped organize several conferences throughout the years.

The presence of Ian and Marty in both the conference and summer school amounted to three weeks during which they were fully available to the participants' questions and expositions of mathematics. This is one of their major contributions to mathematics: it is not just that their list of publications is long and of excellent quality, providing considerable advances in the field, but that so many mathematicians have profited from personally talking to Marty and Ian. The impact of Ian and Marty in mathematics may also be perceived when we see that they have had a total of 56 PhD students and research associates.

Marty and Ian have delivered hundreds of invited talks in places as far apart as Iceland, China, Brazil, Singapore, Australia, Canada and Japan as well as in the United States of America, the United Kingdom and several other european countries. In all countries they have been invited by the most prestigious mathematical institutions.

The recognition of their work is widespread and has awarded them several honours such as the aforementioned Ferran Sunyer i Balaguer Prize, the Farfel Award (University of Houston – MG) and the Michael Faraday Prize (Royal Society – IS).

Finally, mathematics as a whole, and not just the fields of mathematics that attracted Ian and Marty's attention, has seen its image improved through popular and non-technical books signed by Ian and/or Marty. These have probably made many a mathematician's social life considerably less painful.

The papers appearing in these Proceedings reflect well the scope of the work of Martin Golubitsky and Ian Stewart. They illustrate their point of view – symmetry as a form of organizing knowledge, symmetry as means of obtaining model-independent results. Most of the papers include results by one or both in the bibliography; the ones that do not, rely on results based on work by Marty and/or Ian – second generation results, they might be called.

Stewart, Elmhirst and Cohen address a central problem in evolutionary biology, that of speciation. They adopt the viewpoint that speciation is driven by natural selection acting on organisms, with the role of the genes being secondary. Their methods include numerical simulations and analytic techniques from equivariant bifurcation theory, and the conclusions are related to field observations of various organisms. The models are biologically interpreted as speciation appearing as an emergent property at the organism level.

Chillingworth and Golubitsky consider the Landau-de Gennes model for the free energy of a liquid crystal. They discuss the geometry of its equilibrium set for spatially homogeneous states. Using equivariant bifurcation theory they classify square and hexagonally periodic patterns that arise when a nematic state becomes unstable.

The relation between symmetric attractors and ergodic theory is the main subject of Field's paper. It describes recent analytic results on the coexistence of symmetry and chaotic dynamics in equivariant dynamics, studying the important phenomenon of persistent ergodicity and emphasizing the case of stably SRB attractors.

Nicol, Sidorov and Broomhead deal also with questions from ergodic theory, although from a different angle. They consider an IFS (iterated function system) which contracts-on-average. They describe the relation between the Hausdorff dimension of the stationary invariant measure, the entropy and the Lyapunov exponent and the semigroup generated by the functions which define the IFS.

Pattern formation in physical systems is one of the major research themes in mathematics. Quoting Golubitsky and Stewart: "Many instances of pattern formation can be understood within a single framework: the viewpoint of symmetry". Examples of this include the following papers.

Ashwin's paper is concerned with the asymptotic behaviour of evolving patterns on unbounded domains. It discusses appropriate topologies for studying this class of problems, and for overcoming the lack of compactness inherent in them. It develops concepts of attraction and stability for evolution equations on unbounded domains, without reference to spectral properties of the equations. The concepts are illustrated with a discussion of evolving spirals.

Symmetry of attractors is also treated by Goetz and Mendes in the context of Euclidean piecewise rotations, which in a sense constitute the most basic two-dimensional generalizations of interval exchange maps. A new example of attractor is illustrated.

A model for radially forced thermoconvection of a fluid in an annulus based on the two-dimensional Boussinesq fluid equations is considered by Rusu and Langford. They use $O(2)$-equivariant bifurcation theory to classify steady-state patterns and rotating waves corresponding to spatio-temporal vortex patterns for the thermoconvection problem.

Callahan uses equivariant bifurcation theory to study three-dimensional pattern formation by Hopf bifurcation in a system that is homogeneous and isotropic in all three directions, such as a reaction-diffusion system. Callahan restricts attention to solutions that have the periodicity of the face-centered or body-centered cubic lattices. This is one more example where the symmetries of the system are used to predict mechanisms of pattern-formation that are model-independent.

Rucklidge, Silber and Fineberg present an interesting interplay of theory and experiments. The main point is to explain in terms of symmetry and secondary bifurcation three experimentally observed patterns. Two of them can be obtained from the bifurcation and symmetry analysis; for the third the method provides two possible group representations and the Fourier spectrum has to be used to decide between them.

Alonso, Net and Sanchéz present, in terms of $O(2)$ symmetry, numerical results on a Boussinesq fluid, for a low Prandtl number, in the case of 1:2 resonance. Results are presented mainly in the form of figures. It also includes an explanation of the distinction between $O(2)$ annular symmetry and the periodic planar Bénard convection, where there is an additional reflectional symmetry that reduces the effect of resonance.

The onset of convection in systems that are heated via current dissipation in the lower boundary or that lose heat from the top boundary via Newton's law of cooling is posed as a bifurcation problem by Prat, Mercader and Knobloch. They reformulate the convection problem in order to define a bifurcation parameter that remains constant under fixed external conditions. The solutions obtained are then compared with the ones obtained via the standard formulation.

Bayliss, Matkowsky and Aldushin report interesting numerical results on a model for flame propagation in a cylinder, considering the modes known as bound states or asymmetric travelling waves.

Glendinning's paper deals with an interesting and original case within the extremely rich world of "tent maps". Despite its conciseness, the paper contains a stimulating and careful survey of some properties of this special map, which shows its relevance in the analysis of the phenomenon of transition to chaos occurring via a sequence of bifurcations.

Diekmann and van Gils explore the dynamics of age structured populations with a single reproductive age class. They begin by obtaining a symmetric formulation for the projection matrix in this case, and show the existence of invariant manifolds in the equal-sensitivity case.

Cigogna's paper describes recent results on properties of Poincaré-Dulac normal forms and their normalizing transformations for bifurcation problems with resonances. It also discusses how, by imposing that the formal change of coordinates into normal forms is convergent, the normal forms can be used to find multiple periodic solutions.

The more classical work on bifurcation, either using symmetry or not, appears in the papers by Cox and Matthews, where systems with Galilean and Euclidean symmetry are studied, and in the paper by Georgescu, Rocsoreanu and Giurgiteanu, concerning the FitzHugh-Nagumo equations.

The papers now presented in this Proceedings volume, despite their width and breadth, do not necessarily describe all the scientific contributions to the Conference. Some of the invited lecturers did not present a paper for these Proceedings, in most cases because their lecture covered results which had already been submitted for publication elsewhere. This was the case of P. Chossat, B. Dionne, B. Fiedler, E. Knobloch, M. Krupa, J. Lamb, I. Melbourne, M. Roberts and H. Swinney.

The Editors would like to acknowledge the various institutions which partially contributed to the support of the Conference and of the publication of these Proceedings, either through funding or logistics. So we take the opportunity of explicitly thanking: CIM (Centro Internacional de Matemática), CMAUP (Centro de Matemática Aplicada da Universidade do Porto), CAMGSD (Centro de Análise Matemática, Geometria e Sistemas Dinâmicos of Instituto Superior Técnico), Fundação Calouste Gulbenkian, FCT (Fundação para a Ciência e Tecnologia), FLAD (Fundação Luso-Americana para o Desenvolvimento), University of Porto, directly and through the Faculties of Economics (FEP) and Sciences (FCUP), particularly Departamentos de Matemática Pura and Matemática Aplicada. Most of the editorial work done by SC took place at the Mathematics Department of the University of Aarhus, Denmark, whose hospitality is gratefully acknowledged. JB would also like to acknowledge the invaluable support of Prof. João Palhoto de Matos, of the Departamento de Matemática of Instituto Superior Técnico, whose technical expertise helped enormously in the process of bringing these Proceedings to print.

The Editors,   *J. Buescu*
                                                                S.B.S.D. Castro
                                                                A.P.S. Dias
                                                                I.S. Labouriau

Conference on

"Bifurcations, Symmetry and Patterns"

In honour of Martin Golubitsky and Ian Stewart

29 June–4 July, 2000
University of Porto, Porto, Portugal

Organizing Committee

Jorge **Buescu**
Centro de Análise Matemática,
Geometria e Sistemas Dinâmicos
Departamento de Matemática
Instituto Superior Técnico
Av. Rovisco Pais
1049-001 Lisboa, Portugal
jbuescu@math.ist.utl.pt

Ana Paula **Dias**
Centro de Matemática
Departamento de Matemática Pura
Universidade do Porto
Praça Gomes Teixeira
Faculdade de Ciências
Universidade do Porto
4099-002 Porto, Portugal
apdias@fc.up.pt

Sofia **Castro**
Centro de Matemática Aplicada
Faculdade de Economia
Universidade do Porto
Rua Dr. Roberto Frias
4200 Porto, Portugal
sdcastro@fep.up.pt

Isabel **Labouriau**
Centro de Matemática Aplicada
Departamento de Matemática Aplicada
Rua das Taipas, 135
4050-600 Porto, Portugal
islabour@fc.up.pt

Invited Speakers

Peter **Ashwin**
School of Math Sciences
University of Exeter
Exeter EX4 4QE, UK
PAshwin@maths.ex.ac.uk

Pascal **Chossat**
INLN
1361 route des Lucioles
Sophia Antipolis
06560 Valbonne, France
chossat@inln.cnrs.fr

Benoit **Dionne**
Department of Mathematics
and Statistics
University of Ottawa
585 King Edward Avenue
Ottawa
Ontario K1N 6N5, Canada
benoit@mathstat.uottawa.ca

Bernold **Fiedler**
Freie Universität Berlin
Fachbereich Mathematik
und Informatik
Arnimallee 2–6
D-14195 Berlin, Germany
fiedler@math.fu-berlin.de

Michael **Field**
Department of Mathematics
University of Houston
Houston, TX 77204-3476, USA
mf@uh.edu

Martin **Golubitsky**
Department of Mathematics
University of Houston
Houston, TX 77204-3476, USA
mg@uh.edu

Edgar **Knobloch**
Department of Physics
University of California
Berkeley, CA 94720-7300, USA
knobloch@physics.berkeley.edu

Martin **Krupa**
Department of Mathematical Sciences
New Mexico State University
Las Cruces, NM 88003-8001, USA
mkrupa@nmsu.edu

Jeroen **Lamb**
Department of Mathematics
Imperial College
London, SW7 2BZ, UK
jsw.lamb@ic.ac.uk

William **Langford**
Mathematics
and Statistics Department
University of Guelph
Guelph Ontario Canada
N1G 2W1, Canada
wlangfor@msnet.mathstat.uoguelph.ca

Ian **Melbourne**
Department of Mathematics
and Statistics
University of Surrey
Guildford, GU2 5XH, UK
I.Melbourne@surrey.ac.uk

Mark **Roberts**
Department of Mathematics
and Statistics
University of Surrey
Guildford, GU2 5XH, UK
M.Roberts@surrey.ac.uk

Mary **Silber**
Engineering Sciences
and Applied Mathematics
Northwestern University
2145 Sheridan Road
Evanston, IL 60208, USA
m-silber@nwu.edu

Ian **Stewart**
Mathematics Institute
University of Warwick
Coventry CV4 7AL, UK
ins@maths.warwick.ac.uk

Harry **Swinney**
Physics Department and
Center for Nonlinear Dynamics
University of Texas at Austin
Austin, Texas 78712, USA
swinney@chaos.ph.utexas.edu

Invited Lectures

Trends in Mathematics:
Bifurcations, Symmetry and Patterns, 3–54
© 2003 Birkhäuser Verlag Basel/Switzerland

Symmetry-Breaking as an Origin of Species

Ian Stewart, Toby Elmhirst, and Jack Cohen

Abstract. A central problem in evolutionary biology is the occurrence in the fossil record of new species of organisms. Darwin's view, in *The Origin of Species*, was that speciation is the result of gradual accumulations of changes in body-plan and behaviour. Mayr asked why gene-flow failed to prevent speciation, and his answer was the classical allopatric theory in which a small founder population becomes geographically isolated and evolves independently of the main group.

An alternative class of mechanisms, sympatric speciation, assumes that no such isolation occurs. These mechanisms overcome the stabilising effect of gene-flow by invoking selection effects, for example sexual selection and. assortative mating. We interpret sympatric speciation as a form of symmetry-breaking bifurcation, and model it by a system of nonlinear ODEs that is 'all-to-all coupled', that is, equivariant under the action of the symmetric group S_N. We show that such bifurcations can be interpreted as speciation events in which the dominant long-term behaviour is divergence into two species. Generically this divergence occurs by jump bifurcation – 'punctuated equilibrium' in the terminology of evolutionary biology. Despite the discontinuity of such a bifurcation, mean phenotypes change smoothly during such a speciation event. So, arguably, do mean-field genotypes related to continous characters.

Our viewpoint is that speciation is driven by natural selection acting on organisms, with the role of the genes being secondary: to ensure plasticity of phenotypes. This view is supported, for example, by the evolutionary history of African lake cichlids, where over 400 species (with less genetic diversity than humans) have arisen over a period of perhaps 200,000 years. Sympatric speciation of the kind we discuss is invisible to classical mean-field genetics, because mean-field genotypes vary smoothly.

Our methods include numerical simulations and analytic techniques from equivariant bifurcation theory. We focus on two main models: the generic cubic-order truncation of a symmetry-breaking bifurcation in an S_N-equivariant system of ODEs, and the BirdSym system introduced by Elmhirst in which the biological interpretation of variables is more explicit.

We relate our conclusions to field observations of various organisms, including Darwin's finches. We also offer a biological interpretation of our models, in which speciation is represented as an emergent property of a complex system of entities at the organism level. We briefly review questions about selection at the level of groups or species in the light of this interpretation.

'A broken or interrupted range may often be accounted for by the extinction of species in the intermediate regions.'
Charles Darwin, *The Origin of Species*

'Unnatural selection was a fact, but the wizards knew, they *knew*, that you couldn't start off with bananas and get fish.'
Terry Pratchett, Ian Stewart, and Jack Cohen, *The Science of Discworld*

1. Introduction

How do new species arise in evolution? Well before Darwin, it was known that animals and plants could be persuaded to change, in small ways, by the artificial application of selective pressure. Breeding techniques could produce bigger or smaller dogs, redder roses, more nutritious cereal crops. However, no amount of selective breeding seemed able to persuade an organism to change species, let alone generate an entirely new species. Darwin's revolutionary insight was to realise that nature provides its own source of selective pressure – competition for survival. He assembled a wealth of evidence for the 'mutability of species', without human intervention and over long periods of time. However, despite the title of his magnum opus *The Origin of Species*, he did not propose any detailed mechanism by which entirely new species could arise, other than the slow accumulation of small changes.

One of the most persuasive arguments against gradual drift as a cause of speciation is the fact that (sexual) organisms breed. For example, consider the divergence of chimpanzees and early hominids from a common ancestral species about 5 Myr ago. Somehow, a single species became two. On the basis of Darwin's proposal, gradual transition, a difficulty arises. As the two incipient species begin to diverge, the initial changes are so small that the organisms can still interbreed. Exchange of genetic material – what Mayr [57, 58] calls *gene-flow* – will cancel out that small divergence as their descendants 'regress to the mean', Galton [33]. It seems that speciation by gradual changes is a non-starter.

Mayr's answer to this dilemma, building on work of Dobzhansky [25], was the mechanism of *allopatric* speciation. Here some (small) 'founder population' becomes isolated from the main group, perhaps by migration or geographical accident. Once separated, the new group evolves independently of the main one – gene-flow between the groups is switched off – until eventually the two groups are no longer able to interbreed even if they are brought back into contact. When and if they come together again, the two groups will remain separate species: gene-flow between them will have ceased, permanently.

Attractive though it may seem, the allopatric theory also has difficulties – for example, the frequency with which the allegedly divergent groups must have managed to reoccupy the same territory after lengthy periods of isolation stretches credulity. An alternative class of mechanisms, collectively known as *sympatric* speciation, propose various methods for maintaining divergence in the presence of gene-flow. In addition, Eldredge and Gould [26] suggest, in their theory of 'punctuated equilibrium', that speciation might be a *sudden* evolutionary event. Ridley [67] chapter 16 gives a good overview of conventional theories of speciation, and Rice and Hostert [65] survey experimental evidence for or against a variety of theories.

Over the last few years the tide of debate has turned, and sympatric speciation is now considered to have been a common – though by no mean universal – route to new species, see for example Higashi *et al.* [42], Kondrashov and Kondrashov [53], Dieckmann and Doebeli [23] and references therein, Pennisi [64], Rundle *et al.* [69], Huey *et al.* [44], and Winker [81]. Behind this change of attitude is a growing realisation that the old gene-flow arguments involve several tacit assumptions: for example that gradually changing causes must produce gradually changing effects, and that uniform behaviour in uniformly changing conditions should remain uniform. Even when made explicit, those assumptions seem plausible; how much more persuasive must they have been when they remained tacit? They are, however, wrong. In the terminology of modern nonlinear dynamics, the first says that jump bifurcation is impossible and the second tells us that symmetries cannot break. In fact, jump bifurcation and symmetry-breaking are generic phenomena in nonlinear dynamical systems.

In this paper we describe recent work, much of it previously unpublished, which applies the viewpoint of nonlinear dynamics to derive a general mathematical setting in which a form of sympatric speciation is both natural and inevitable under appropriate conditions. The ideas apply to an entire class of models, depending on the biological interpretation of the system and its parameters. In order to illustrate these possibilities we describe a more specific model, BirdSym, introduced by Elmhirst [28] in the context of speciation in birds.

This paper is addressed to several audiences, primarily the nonlinear dynamics community (which will find the biology unfamiliar) and the evolutionary biology community (which will find the mathematics unfamiliar). For this reason we shall discuss some basic biological and mathematical background, and illustrate the mathematics with simple examples. Technical details, for the experts, are collected in the appendix. Our main focus is on the mathematical techniques, but we also attempt to relate our models to possible biological mechanisms. The generality of the mathematical phenomena leads us to suspect that sympatric speciation can occur through a wide variety of mechanisms, differing from each other only in fine detail: the search for a *unique* biological mechanism for sympatric speciation is almost certainly misguided. However, the same 'meta-mechanism' probably covers most instances of sympatric speciation, and it can be summarised for a nonlinear dynamics audience in simple terms:

- Speciation in a system of nominally identical organisms occurs when the single-species state loses dynamic stability and no alternative stable single-species state is available.

The biological problems are to understand the nature of such instabilities and to interpret the consequent bifurcations: again there is no reason to expect a unique explanation. Equally, there is no reason to suppose that the particular mathematical context described here is the only reasonable one. On the contrary, we have deliberately idealised the context in order to emphasise a number of phenomena that might otherwise appear counter-intuitive.

The main mathematical conclusions of our models are:

- Sympatric speciation is a generic phenomenon in *nonlinear* systems.
- The most common divergence is to two species, but intermediate stages with three species can occur for some parameter values.
- Speciation occurs through a jump bifurcation.
- Mean phenotypes change continuously throughout the bifurcation.
- In contrast, the variance of the equilibrium phenotype across the whole population changes discontinuously at the bifurcation point.

Our main biological conclusions are:

- Sympatric speciation is driven by the interactions of organisms and their environment, especially other organisms.
- Sympatric speciation occurs when a 'generalist' strategy for exploiting the environment becomes less successful than a 'specialist' division of labour.
- The role of genes is to render phenotypes plastic: genetics affects speciation only indirectly.
- The important plasticity results from recombination, not from mutation.
- Speciation involves no change to the frequency with which any particular allele occurs in the mean-field gene pool: instead, what changes is how complexes of alleles are associated in organisms that survive to breed.

2. Speciation Mechanisms in Biology

About 5 Myr ago a small group of bedraggled finches was blown by a storm to an isolated Pacific archipelago, the Galápagos Islands. The new arrivals found few predators and their main competition was sea-birds; they prospered. As the finch population grew, it diversified. When Darwin arrived at the islands on board the *Beagle* in 1837-38 he found 13 distinct species of finch. A fourteenth, found in the Cocos Islands, has since been added to the list. Collectively, these 14 species are known as 'Darwin's finches', Fig.1. See Lack [54], Grant [37].

At first sight, Darwin's finches are a classic example of allopatric speciation: a small founder population plus geographical isolation. All that is missing (and it is not essential) is the eventual reunification with the original source.

However, some features of Darwin's finches do not fit the allopatric model so convincingly. In particular, their diversification occurred *after* their arrival at the new habitat. The Galápagos Islands are tiny, and close enough for a finch to fly from one to another. So why did the population split? There is a term for such behaviour: *radiation*. A large enough change in environment can trigger a burst

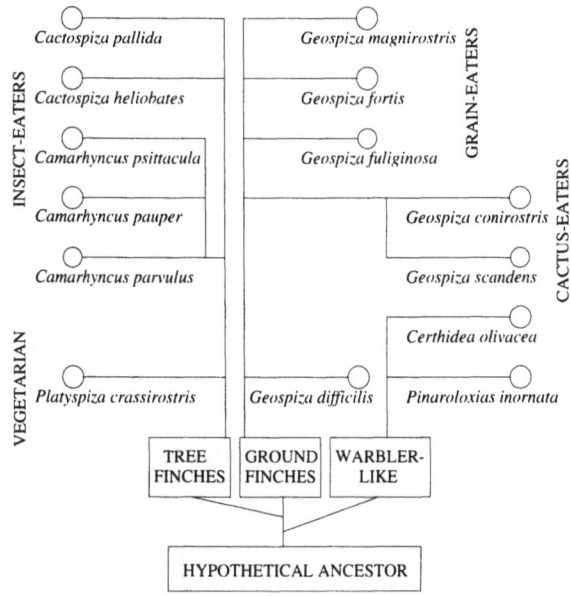

FIGURE 1. Phylogeny of Darwin's finches

of diversity as organisms explore a wide array of potential environmental niches. Different organisms from the same species specialise on different features of the environment.

However, as just stated, the Galápagos Islands are tiny, and close enough for a finch to fly from one to another. This implies that the *potential* environment – or range of environments – is the same for all finches. In other words, radiation is closer to sympatric speciation than it is to allopatric speciation. 'Radiation' is more a description than an explanation.

Speciation is complicated because a wide range of distinct factors could be – in some sense must be – involved. These factors can be split into at least three broad classes:

- Genotype
- Phenotype
- Environment

Genotype (genetic makeup) and phenotype (bodily form and behaviour) differ from one bird to another. They are related, but not trivially so – forget simplified images of genes being the 'blueprint' for an organism. Many genetic changes are cryptic, with no obvious effect – indeed no effect – on phenotype. Many phenotypic changes have no obvious – indeed no – genetic cause. See Cohen and Stewart [15].

'Environment' is an even broader term. An organism's environment may vary both spatially (geography) and temporally (history). Even in a limited local region, many different environmental strategies may be available – hide in shade, bask in sunshine, exploit the top/middle/bottom of a tree, eat insects/seeds/cactus. And environmental factors can include climate, weather, terrain, and other organisms – especially those of the organism's own species, its main competitors. Disentangling all of these possibilities and their interconnections is difficult, perhaps impossible; not surprisingly, the whole area is riddled with controversy.

The mathematician's reflex response to a complex system of interactions is not to catalogue them, but to simplify them. Many factors are omitted, others are replaced by idealised versions. This process is easily misunderstood. Its aim is not to impose such artificial simplicity on the complexities of reality. Instead, it is a way of exploring which features of the system are important, by removing details and seeing what happens. The test of a mathematical model does not lie in the accuracy of its assumptions, but in the accuracy of its conclusions.

In this spirit, we focus on a way to model speciation that may illuminate some aspects of that process. The most relevant question, for the purposes of this paper, is:

- Is it reasonable for a single species to split into two or more species, on the assumptions that all individuals are potentially able to interbreed (panmixis) and that at any instant all organisms are exposed to the same environment as the others (sympatry)?

Thus we ask whether selective breeding patterns or diverse geography are actually *necessary* for speciation – not whether they occur. We will show that they are not necessary, by exhibiting a general class of processes in which speciation can occur in a panmictic sympatric population. We do not claim that all speciation occurs in such a manner: merely that it *can*. Notice that our simplifying assumptions tend to strengthen the message, not weaken it: we remove two of the most obvious sources of species divergence and show that even in their absence, such divergence can still occur. The widely held intuition that selective breeding or geographical discontinuities are *necessary* for speciation is thereby challenged. Nonetheless, if either factor *is* present, it may render speciation more likely.

3. Bifurcation and Symmetry-Breaking

Until fairly recently – the main trend set in around 1960 and became full-blown by about 1980 – most mathematical modelling employed linear equations. There were glorious exceptions, such as celestial mechanics, shock waves, and the Hodgkin-Huxley equations for the nerve impulse, but in most areas of applied science the straight line graph ruled. In many areas (sociology springs to mind, and educational theory) it still does.

Linear models were employed for two good reasons:

- They are simple to describe and relatively easy to analyse
- They capture many phenomena adequately and thereby aid understanding

On the other hand, it turns out that the world of nonlinear phenomena is far richer than the linear world, and that most of these phenomena seem counterintuitive to a mind raised on linear mathematics. This might not matter, except that in most respects nature seems to behave nonlinearly.

Many authors model speciation in terms of dynamical systems (systems of ODEs), and we shall follow their lead. See for instance Kawecki [50], Dieckmann and Doebeli [23], and Hofbauer and Sigmund [43]. We introduce a class of nonlinear models that will allow us to examine the effects of nonlinearity in the speciation process.

Bifurcation

One characteristic feature of nonlinear dynamics is *bifurcation* – a rapid change in effect brought about by a small change in cause. The next example illustrates this phenomenon in a simple model.

Example 1. Let $x \in \mathbf{R}$ represent the state of some system, let $\lambda \in \mathbf{R}$ be a parameter, and suppose that the time evolution of the system is determined by the ODE

$$dx/dt = \lambda x + 2x^2 - x^3$$

The dynamics of this system is illustrated in Fig. 2. The solid and dotted lines show how the equilibria $dx/dt = 0$ vary with λ: a solid line indicates a stable equilibrium, a dotted line an unstable one. The arrows show the direction in which x changes when the system is not in equilibrium. Two values of λ (namely -1 and 0) are special: qualitative changes in the dynamics occur as λ varies through them.

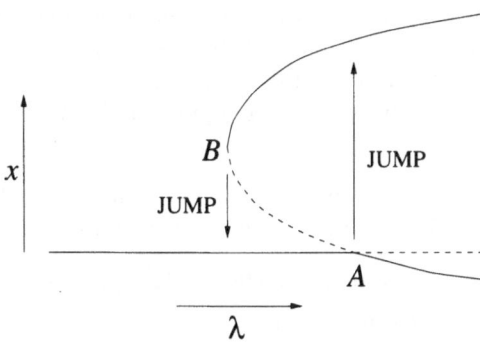

FIGURE 2. Bifurcation diagram illustrating jumps and hysteresis.

Point A, where $\lambda = 0$, is a point of *transcritical* bifurcation. For $\lambda < 0$ the equilibrium $x = 0$ is stable, but it becomes unstable for $\lambda > 0$. Meanwhile a previously unstable 'branch' – a segment of the parabola $\lambda = x^2 - 2x$ between points B and A – becomes stable for $\lambda > 0$. There is an 'exchange of stability' between the two branches.

Point B, where $\lambda = -1$, is a different kind of bifurcation, a *fold* or *saddle-node* point, see Arrowsmith and Place [1]. For $\lambda < -1$ there are no equilibria near B (only the distant one at $x = 0$). As λ increases through -1, two new branches – one stable, the other unstable – separate from B.

Suppose the system starts out with $\lambda < -1$, and let λ slowly increase. We assume *quasi-static* variation, in which x settles down to the closest available equilibrium much faster than λ is changing. The system therefore remains at $x = 0$ when λ becomes greater than -1, even though a new stable equilibrium has appeared. However, as λ passes through 0, the value of x suddenly changes to the only available stable equilibrium, on the top branch that emanates from B. This is a *jump bifurcation*, and we see that when λ makes a very small change through -1, the value of x changes substantially. Indeed, in an idealisation where the timescale for λ is infinitely slow compared to that for x, we see that a continuous change in λ causes a discontinuous change in x.

In short: small changes in the cause can sometimes have big effects. Not always: here, if $\lambda \neq -1, 0$ then small changes in the cause have small effects. Big effects are rare, but can still be unavoidable as a parameter varies.

Suppose that after the jump has occurred, we decrease λ back below 0. Because the top arc of the parabola is stable for $\lambda > -1$, the system does not reverse its previous change and jump back to $x = 0$ when λ passes through 0. Instead, it remains on the top branch until λ gets below -1. Only then does it jump back to $x = 0$. We therefore observe 'irreversible' effects as the parameter varies – the phenomenon of *hysteresis*.

Symmetry-Breaking

Bifurcation, jumps, and hysteresis are typical in nonlinear dynamical systems, along with considerably more complex behaviour. Linear systems display bifurcation only in a very trivial way (a stable equilibrium suddenly becomes unstable). Because a linear system always has a unique equilibrium, it cannot display hysteresis.

The other key phenomenon for this paper is *symmetry-breaking*: here a symmetric system of equations can have solutions with *less* symmetry, or none.

Example 2. Consider the system of ODEs on $\mathbf{R}^3$ given by

$$\begin{aligned} dx/dt &= \lambda x - (x + y + z) + x^2 \\ dy/dt &= \lambda y - (x + y + z) + y^2 \\ dz/dt &= \lambda z - (x + y + z) + z^2 \end{aligned} \tag{1}$$

This system is symmetric under the group $\mathbf{S}_3$ of all permutations of (x, y, z). There is a trivial solution $x = y = z = 0$ and the linearisation about this solution has matrix

$$L = \begin{bmatrix} \lambda - 1 & -1 & -1 \\ -1 & \lambda - 1 & -1 \\ -1 & -1 & \lambda - 1 \end{bmatrix}$$

with eigenvectors and eigenvalues as follows:

$$
\begin{aligned}
v_0 &= (1,1,1)^T : \text{eigenvalue } \lambda - 3 \\
v_1 &= (1,-1,0)^T : \text{eigenvalue } \lambda \\
v_2 &= (0,1,-1)^T : \text{eigenvalue } \lambda
\end{aligned}
$$

Here the T indicates the transpose, to avoid printing column vectors.

When $\lambda < 0$ the trivial solution is stable (all eigenvalues negative) but it loses stability when $\lambda > 0$. The kernel of L is two-dimensional when $\lambda = 0$, spanned by v_1, v_2. Note that $v_1 + v_2 = (1,0,-1)^T$ and $\mathbf{R}\{v_1, v_2\} = \{v : x + y + z = 0\}$, the nontrivial irreducible representation of $\mathbf{S}_3$.

We look for equilibria of (1) with λ and x near 0. Then

$$
\lambda x + x^2 = \lambda y + y^2 = \lambda z + z^2 = x + y + z
$$

It is easy to show that near $(\lambda, x) = (0,0)$ these equations imply that at least two of x, y, z are equal, and that all three are equal only when $x = y = z = 0$. (However, there is another bifurcation at $\lambda = 3$ where the kernel of L is the trivial representation of $\mathbf{S}_3$, and a new branch with $x = y = z \neq 0$ bifurcates there.)

By symmetry we may assume $x = y$ and $z \neq x$. Then

$$
\lambda x + x^2 = \lambda z + z^2 \tag{2}
$$

$$
\lambda x + x^2 = 2x + z \tag{3}
$$

Equation (2) implies that $\lambda(x-z) + (x+z)(x-z) = 0$, so that $x + z = -\lambda$, whence $z = -x - \lambda$. Then (3) implies that $\lambda x + x^2 = x - \lambda$. Solving the quadratic equation and retaining only the solution near the origin, we find that a nontrivial solution

$$
\begin{aligned}
x = y &= \tfrac{-\lambda + 1 - \sqrt{\lambda^2 - 6\lambda + 1}}{2} \\
z &= \tfrac{-\lambda - 1 + \sqrt{\lambda^2 - 6\lambda + 1}}{2}
\end{aligned} \tag{4}
$$

exists for all λ near 0. By symmetry, two other 'conjugate' branches also exist, in which x, y, z are permuted.

We investigate the geometry of these new solutions. For $\lambda \sim 0$ we have

$$
\sqrt{\lambda^2 - 6\lambda + 1} \sim 1 - 3\lambda - 4\lambda^2 + O(\lambda^3)
$$

by the binomial theorem, so

$$
\begin{aligned}
x &= \lambda + 2\lambda^2 + O(\lambda^3) \\
y &= \lambda + 2\lambda^2 + O(\lambda^3) \\
z &= -2\lambda - 2\lambda^2 + O(\lambda^3)
\end{aligned}
$$

This is a transcritical branch, existing for all λ near 0. There are two other branches obtained by permuting (x, y, z). Schematically, the bifurcation diagram looks like Fig. 3.

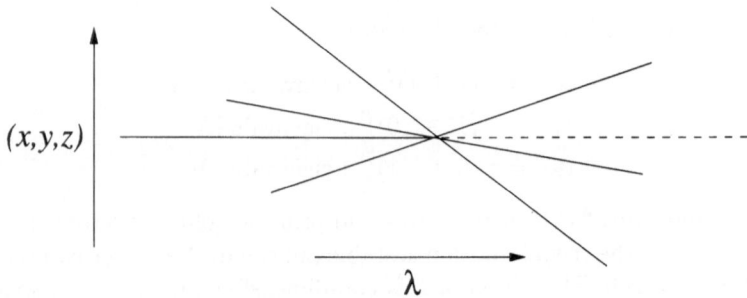

(x,y,z)

λ

FIGURE 3. Bifurcation diagram for symmetry-breaking example.

What about stability? We now show that the above primary branch, and therefore also its conjugates, are *unstable* both before and after bifurcation. The Jacobian is

$$J = \begin{bmatrix} \lambda - 1 + 2x & -1 & -1 \\ -1 & \lambda - 1 + 2y & -1 \\ -1 & -1 & \lambda - 1 + 2z \end{bmatrix}$$

One eigenvector is $(1, -1, 0)^T$ with eigenvalue

$$\varepsilon_1 = \lambda + 2x$$

(we here use the fact that $x = y$). We compute the other two eigenvalues by defining

$$u = (1, 1, 0)^T$$
$$v = (0, 0, 1)^T$$

We have

$$Ju = (\lambda - 2 + 2x)u - 2v, \quad Jv = -u + (\lambda - 1 + 2z)v$$

Thus $\mathbf{R}\{u, v\}$ is invariant under J. On this space the matrix of J is

$$K = \begin{bmatrix} \lambda - 2 + 2x & -2 \\ -1 & -\lambda - 1 - 2x \end{bmatrix}$$

where we have replaced z by $-x - \lambda$. The trace of K is

$$T = -3$$

and the determinant is

$$D = -(\lambda + 2x - 2)(\lambda + 2x + 1) - 2$$

Therefore the eigenvalues of K are

$$\varepsilon_2 = \frac{-3 + \sqrt{9 - 4D}}{2}$$

$$\varepsilon_3 = \frac{-3 - \sqrt{9 - 4D}}{2}$$

When λ is small, $\varepsilon_2 \sim 0$ and $\varepsilon_3 \sim -3$. Computing to first order in λ, we find

$$\begin{aligned}
\varepsilon_1 &= 3\lambda + O(\lambda^2) \\
\varepsilon_2 &= -\lambda + O(\lambda^2) \\
\varepsilon_3 &= -3 + \lambda + O(\lambda^2)
\end{aligned}$$

Near the origin, ε_1 and ε_2 have opposite signs, so the solution branch is unstable near the origin. By symmetry, the same applies to the two conjugate branches. So locally there are no nontrivial stable branches other than the origin, and the origin is stable only when $\lambda < 0$.

Thus, although the equations (1) of Example 2 have branches of equilibria corresponding to two species, those branches are unstable. As it stands, this is a highly unsatisfactory conclusion: we want the speciation process to lead to stable branches. To make matters worse, it turns out that the same difficulty arises for $\mathbf{S}_N$-symmetric models: generically, all primary branches of equilibria are unstable near the origin. That is, the classic 'exchange of stability' does not occur (later we explain why: symmetry is the cause).

Another feature of the calculation deserves attention. Along each bifurcating branch,

$$x + y + z = 2\lambda^2 + O(\lambda^3)$$

which is quadratic in λ. Therefore the branches are all tangent to the plane $x + y + z = 0$ at the origin. Moreover, to first order in λ we have $x + y + z = 0$ on all branches, including the trivial branch, so the mean $\frac{x+y+z}{3}$ is approximately constant (the constant being zero) throughout the bifurcation.

Fold Points and Secondary Branches

We return to the instability of the branches in Example 2. This deficiency of the model is also visible in Example 1, where the branches near the origin are unstable. However, in that example the presence of a cubic term in the ODE causes a primary branch to 'turn round' at a fold point, where it regains stability. We therefore expect to rectify the instability of primary branches by including cubic terms in the model, as well as linear and quadratic.

Experience with the case $N = 3$, which is well understood (see Golubitsky and Schaeffer [34] and Golubitsky *et al.* [36] Chapter XV §4) reinforces this expectation. The group $\mathbf{S}_3 \cong \mathbf{D}_3$ and the relevant action is the standard action of $\mathbf{D}_3$ on $\mathbf{R}^2$. Singularity-theoretic methods reduce the problem to the normal form

$$F\begin{bmatrix} x \\ y \end{bmatrix} = \begin{bmatrix} x(x^2 + y^2 - \lambda) + (x^2 - y^2)(x^2 + y^2 + \mu(x^3 - 3xy^2)) - \alpha) \\ y(x^2 + y^2 - \lambda) + 2xy(x^2 + y^2 + \mu(x^3 - 3xy^2)) - \alpha) \end{bmatrix}$$

where α, μ are parameters and λ is the bifurcation parameter. For purposes of illustration, we assume that this normal form has been derived from a model whose variables are population sizes, so that the branches can be interpreted in terms of species. Depending on the signs of α, μ, there can exist secondary branches with trivial symmetry (that is, three-species branches). The geometry of the bifurcation diagrams is shown in Figure 4. When $\alpha < 0, \mu > 0$ the three-species branch can be stable, but it exists only for a finite interval of values of λ.

Notice that the presence of cubic terms in the normal form here causes the primary branch to turn round at a fold point, and (in this case) it regains stability there. It can also *lose* stability by encountering a secondary branch, or regain stability by encountering a secondary branch, depending on delicate features of the equations.

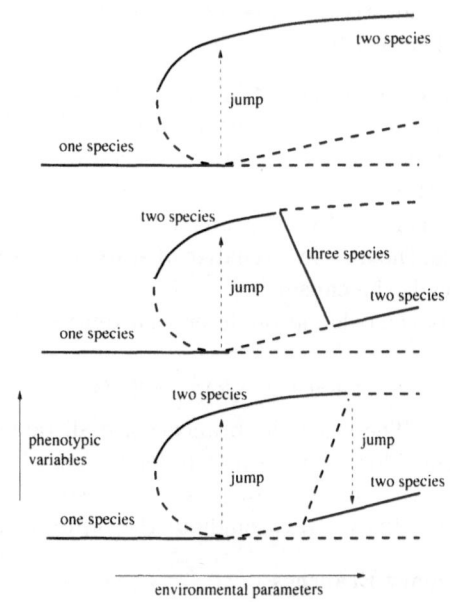

FIGURE 4. Existence of 3-species secondary bifurcations in the case $N = 3$. (Top) $\alpha > 0$. (Middle) $\alpha < 0, \mu > 0$. (Bottom) $\alpha < 0, \mu < 0$.

We show later that similar (but slightly more complicated) features occur in S_N-equivariant bifurcation problems for $N > 3$, with interesting implications for our model of sympatric speciation.

4. Methods of Equivariant Bifurcation Theory

The general context for a symmetry-based analysis of pattern formation in equivariant dynamical systems is symmetric (or equivariant) bifurcation theory. This is surveyed in Golubitsky *et al.* [36]. We briefly summarise the main ideas and state the basic existence and stability theorems for steady-state bifurcation.

Let Γ be a Lie group of linear transformations of $\mathbf{R}^n$. We say that f is Γ-*equivariant* if

$$f(\gamma x, \lambda) = \gamma f(x, \lambda) \tag{5}$$

for all $\gamma \in \Gamma$. Consider a Γ-equivariant ODE

$$\frac{dx}{dt} + f(x, \lambda) = 0 \tag{6}$$

where $x \in \mathbf{R}^n$, $\lambda \in \mathbf{R}$. Steady states are zeros of f. The symmetry of f implies that zeros come in symmetrically related sets. Specifically, if $x \in \mathbf{R}^n$ then the *orbit* of x under Γ is

$$\Gamma x = \{\gamma x : \gamma \in \Gamma\}.$$

If $f : V \to V$ is Γ-equivariant, then the zero-set of f is a union of Γ-orbits, for if $f(x) = 0$ then $f(\gamma x) = \gamma f(x) = \gamma 0 = 0$. It is convenient to consider solutions in the same Γ-orbit as being 'the same solution' up to symmetry.

The *isotropy subgroup* of $x \in V$ is

$$\Sigma_x = \{\sigma \in \Gamma : \sigma x = x\}.$$

Isotropy subgroups of points on the same orbit are conjugate. Indeed $\Sigma_{\gamma x} = \gamma \Sigma_x \gamma^{-1}$. We therefore tend not to distinguish between isotropy subgroups and their conjugates. The *isotropy lattice* is the partially ordered set formed by all conjugacy classes of isotropy subgroups, with ordering induced by inclusion. It is a finite partially ordered set but, despite its name, it is not always a lattice in the strict algebraic sense.

If $H \subset \Gamma$ is any subgroup, we define the *fixed-point subspace*

$$\mathrm{Fix}(H) = \{x \in V : \gamma x = x \;\; \forall \gamma \in H\}.$$

We have $\mathrm{Fix}(\gamma H \gamma^{-1}) = \gamma \, \mathrm{Fix}(H)$.

For simplicity, assume that $f(0, \lambda) \equiv 0$, so there exists a 'trivial branch' of solutions $x = 0$. The *linearisation* of f is

$$L_\lambda = D_x f|_{0,\lambda}.$$

Local bifurcation at $\lambda = 0$ occurs when the trivial branch undergoes a change of linear stability, so that L_0 has eigenvalues on the imaginary axis (often called *critical eigenvalues*). There are two cases:

- Steady-state bifurcation: L_0 has a zero eigenvalue.
- Hopf bifurcation: L_0 has a complex conjugate pair of purely imaginary eigenvalues.

Hopf bifurcation leads to time-periodic solutions and will not be considered in this paper.

A bifurcating branch *breaks symmetry* if the corresponding isotropy subgroup is not the whole of Γ. We can detect broken symmetry using fixed-point subspaces. A crucial feature of fixed-point subspaces is that they are dynamically invariant: if f is Γ-equivariant and $H \subset \Gamma$ then f leaves $\mathrm{Fix}(H)$ invariant. The proof is trivial, but the result is very useful. Suppose we are seeking a branch of solutions to a Γ-equivariant bifurcation problem $f(x, \lambda) = 0$, breaking symmetry to Σ. Then $x \in \mathrm{Fix}(\Sigma)$, and it suffices to solve $f|_{\mathrm{Fix}(\Sigma)} = 0$. Since $\mathrm{Fix}(\Sigma)$ has smaller dimension than n, in the symmetry-breaking case, this is in principle a simpler problem.

To study local bifurcations, we look at the *critical eigenspace*: the real generalised eigenspace E for the critical eigenvalues. This is the kernel of L_0^n in the steady-state case. By equivariance, E is a Γ-invariant subspace of $\mathbf{R}^n$. The analysis now proceeds in a sequence of stages. First, we determine the generic possibilities for the action of Γ on E. By 'generic' we mean 'unable to be destroyed by a small perturbation of f'. For steady-state bifurcation, generically E is absolutely irreducible, meaning that the only equivariant linear maps are scalar multiples of the identity. Next, we use Liapunov-Schmidt or centre manifold reduction to reduce the problem to one posed on the E. With sensible choices in the reduction procedure, the reduced problem is Γ-equivariant. See Golubitsky *et al.* [36]. Next, we apply the Equivariant Branching Lemma (see below) or more detailed analysis to show existence of symmetry-breaking solutions. Finally, we study the stability of bifurcating solutions.

The Equivariant Branching Lemma is the simplest and most widely used existence theorem for steady-state branches. To state it we require the concept of an *axial* subgroup: this is an isotropy subgroup Σ for which $\dim \text{Fix}(\Sigma) = 1$. For such isotropy subgroups we have the following basic existence theorem of Vanderbauwhede [80] and Cicogna [14]:

Theorem 3. *(Equivariant Branching Lemma). Let $f(x, \lambda) = 0$ be a Γ-equivariant bifurcation problem where $\text{Fix}(\Gamma) = 0$. Let Σ be an axial subgroup. Then generically there exists a branch of solutions to $f(x, \lambda) = 0$ emanating from the origin with symmetry group Σ.*

For a proof see Golubitsky *et al.* [36] Chapter XIII Theorem 3.3 p.82.

Next, we discuss stability of the solutions. A bifurcating steady state (x_0, λ_0) is stable if all eigenvalues of $L_{(x_0, \lambda_0)} = D_x f|_{(x_0, \lambda_0)}$ have real parts < 0, and unstable if at least one eigenvalue has real part > 0. There is one situation in which we can guarantee that all branches predicted by the Equivariant Branching Lemma are generically *unstable*, and this situation arises in the speciation model and influences its analysis and development.

Theorem 4. *Assume that*

- Γ *acts absolutely irreducibly on the critical eigenspace E.*
- $D_x f|_{(0,\lambda)} = c(\lambda)I$ *where $c : \mathbf{R} \to \mathbf{R}$.*
- $c(0) = 0$ *and $c'(0) < 0$.*
- Σ *is an axial subgroup of Γ.*
- *Some term in the Taylor series of $f|_{\text{Fix}(\Sigma) \times \{0\}}$ is nonzero.*
- $(D_x q)_{x_0}$ *has eigenvalues off the imaginary axis, where q is the quadratic part of f and $x_0 \in \text{Fix}(\Sigma)$.*

Then the branch of solutions in $\text{Fix}(\Sigma) \times \mathbf{R}$ guaranteed by the Equivariant Branching Lemma consists of unstable solutions.

This theorem is originally due to Ihrig. For a proof and qualifying remarks, see Golubitsky *et al.* [36] Chapter XIII Theorem 4.4 p.90. The key feature here is the existence of a nonzero quadratic term in the Taylor series of f, which for some group actions is a generic feature. In particular, this is the case for $\mathbf{S}_N$ in its standard absolutely irreducible action on $\mathbf{R}^{N-1}$.

5. Model Equations

In order to model speciation with ODEs we need to discretise the distribution of phenotypes into a fixed number of 'tokens'. These tokens provide a coarse-graining of the population into N clumps, which we call PODs (Placeholders for Organism Dynamics). The choice of N is a modelling convention: something in the range 10 to 100 seems reasonable in practice.

The use of PODs avoids the problem of individual organisms dying and new ones coming into existence. A POD is similar to the existing concepts of a deme (Salthe [70]) or lineage (Rollo [68]). Its biological meaning might, perhaps, be interpreted in the following manner. In field observations, animals in a given species are commonly seen to have different habits. Some range widely, some accept low quality food locally, and so on. These are behavioural qualities that would define PODs subjectively in the field.

Associated with POD i, for $1 \leq i \leq N$, is a phenotypic variable $x_i \in \mathbf{R}^k$ representing a vector of k continuous characters. For purposes of illustration we usually take $k = 1$, but the entire theory goes through to the general case with only minor technical modifications. The value of x_i is interpreted as the average phenotype in POD i.

We have $\mathbf{S}_N$ acting on $\mathbf{R}^N$ by permutations, and we model speciation by a parametrised system of $\mathbf{S}_N$-equivariant ODEs on $\mathbf{R}^{Nk}$:

$$dx_i/dt = F(x_1, \ldots, x_N; a_1, \ldots, a_r)$$

Here $x = (x_1, \ldots, x_N)$ represents the phenotypes of the population of N PODs, and $a = (a_1, \ldots, a_r)$ represents 'environmental' parameters.

Equivariance restricts the form of F considerably. It is well known (see Appendix 1) that polynomial invariants for $\mathbf{S}_N$ on $\mathbf{R}^N$ have a Hilbert basis given by sums of kth powers

$$\pi_k = \sum_{i=1}^{N} x_i^k$$

where $k = 1, \ldots, N$. We define $\mathbf{S}_N$-equivariants E_k, for $k = 0, 1, 2, \ldots$ by

$$E_k = (x_1^k, x_2^k, \ldots, x_N^k)^T \tag{7}$$

We prove in Appendix 1 that the $\mathbf{S}_N$-equivariant polynomial mappings are generated over the $\mathbf{S}_N$-invariant polynomial functions by $E_0, \ldots, E_{N-1}$.

In particular, there are two independent quadratic equivariants, $\pi_1 E_1$ and E_2. This turns out to make Theorem 4 applicable, implying that an adequate model should include terms to at least cubic order. The general cubic-degree $\mathbf{S}_N$-equivariant ODE has 13 parameters:

$$
\begin{aligned}
dx_i/dt &= F_i(x) \\
&= a_0 + b_1\pi_1 + b_2 x_i + c_3\pi_1 x_i + c_4 x_i^2 \\
&\quad + d_1\pi_1^3 + d_2\pi_1\pi_2 + d_3\pi_3 + d_4\pi_1^2 x_i + d_5\pi_2 x_i + d_6 x_i^3
\end{aligned}
\tag{8}
$$

Axial Subgroups

It is easy to compute the isotropy subgroup Σ_x of $\mathbf{S}_N$ acting on $\mathbf{R}^N$, for arbitrary x. If $\sigma \in \mathbf{S}_N$ and $\sigma.x = x$, then the only entries x_i, x_j of x that can be permuted by σ are those with equal values, $x_i = x_j$. We therefore partition $\{1,\ldots,N\}$ into disjoint blocks $B_1,\ldots,B_k$ with the property that $x_i = x_j$ if and only if i,j belong to the same block. Letting $b_\ell = |B_\ell|$, we find that

$$
\Sigma_x = \mathbf{S}_{b_1} \times \cdots \times \mathbf{S}_{b_k}
$$

where $\mathbf{S}_{b_\ell}$ is the symmetric group on block B_ℓ.

Up to conjugacy we may assume that

$$
\begin{aligned}
B_1 &= \{1,\ldots,b_1\} \\
B_2 &= \{b_1+1,\ldots,b_1+b_2\} \\
&\quad\cdots \\
B_k &= \{b_1+\cdots+b_{k-1}+1,\ldots,b_N\}
\end{aligned}
$$

Moreover, we can assume that $b_1 \le b_2 \le \cdots \le b_k$. Therefore conjugacy classes of isotropy subgroups of $\mathbf{S}_N$ are in one-to-one correspondence with partitions of N into nonzero natural numbers arranged in ascending order.

Suppose that Σ is an isotropy subgroup corresponding to the simplest non-trivial partition $P = \{p,q\}$ where $p + q = N$ and $p \le N/2$. It is then easy to see that $\mathrm{Fix}(\Sigma)$ consists of all vectors

$$
(\underbrace{u,\ldots,u}_{p}, \underbrace{v,\ldots,v}_{q})
$$

for real numbers u, v. (Here u and v may be equal, or not.) Therefore $\dim \mathrm{Fix}(\Sigma) = 2$. Similarly, if Σ corresponds to a partition of N into k blocks, then $\dim \mathrm{Fix}(\Sigma) = k$.

Finally, we restrict the action of $\mathbf{S}_N$ onto the standard irreducible $\mathbf{R}^{N-1}$, by imposing the relation $x_1 + \cdots + x_N = 0$. The isotropy subgroups remain the same, but the dimension of every fixed-point subspace is reduced by 1. In particular the isotropy subgroups $\mathbf{S}_p \times \mathbf{S}_q$, where $p+q = N$, are the ones that have 1-dimensional fixed-point subspaces, so these are the axial subgroups. By the Equivariant Branching Lemma, generically there exist branches of equilibria with isotropy subgroups $\mathbf{S}_p \times \mathbf{S}_q$. We call these *primary* branches. Not surprisingly, these subgroups will play a major role from now on.

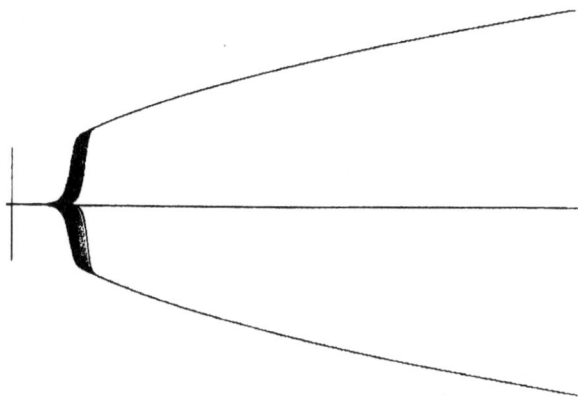

FIGURE 5. Simulation of speciation for 100 PODs.

Preliminary Simulations

Before embarking on an analysis of the bifurcations of (8) we give a preview of the likely behaviour, by numerical simulation. We take $b_2 = \lambda$ as bifurcation parameter, and set $b_1 = -1, c_4 = 1, d_6 = -1$ and all other parameters equal 0. Thus the ODE takes the form:

$$dx_i/dt = \lambda x_i - (x_1 + \cdots + x_N) + x_i^2 - x_i^3 \quad (1 \leq i \leq N)$$

Figure 5 shows a numerical simulation of a solution of these equations for $N = 100$. The picture is obtained by increasing λ by a tiny amount at each step of a numerical integration, while using the values of x_i obtained on the previous step as initial values for the next step. This technique is called *ramping the bifurcation parameter*.

Initially the system is at the origin. There is a relatively rapid jump bifurcation, after which the values of the x_i settle down into two clusters. One cluster contains 48 of the 100 variables x_i, and the common value is positive; the other cluster contains 52 of the variables and the common value is negative. Again we see symmetry-breaking. In fact the isotropy subgroup of the bifurcating branch is $\mathbf{S}_{48} \times \mathbf{S}_{52} \subseteq \mathbf{S}_{100}$.

Figure 6 shows histograms of the distribution of phenotypes in a typical case: note the rapid initial broadening into a highly non-normal distribution, and the subsequent clustering into two peaks, with some intermediate transitions.

Linear Analysis

We begin by carrying out a linear analysis. We normalise the phenotypic variables to represent the deviation from the initial phenotype (single species), which removes the constant term E_0. Thus we assume that prior to bifurcation the phenotype is $x = (0, \ldots, 0)$.

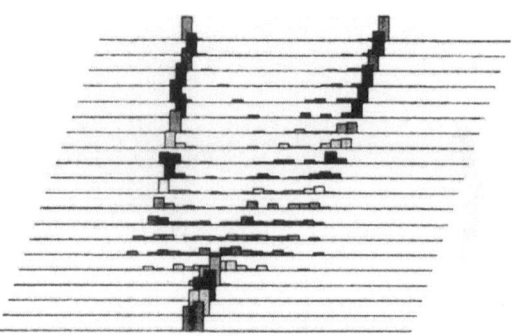

FIGURE 6. Typical sequence of histograms of phenotype distribution during a speciation event.

There are two linear equivariants because $\mathbf{R}^N$ is a direct sum of two absolutely irreducible subspaces V_0, V_1, where

$$V_0 = \mathbf{R}(1,1,\ldots,1)$$
$$V_1 = \{(x_1,\ldots,x_N) : x_1 + \cdots + x_N = 0\}$$

The map E_1 is the identity, and $\sigma_1 E_0 = M$ where

$$M = \begin{bmatrix} 1 & 1 & \cdots & 1 \\ 1 & 1 & \cdots & 1 \\ \vdots & \vdots & \ddots & \vdots \\ 1 & 1 & \cdots & 1 \end{bmatrix}$$

The general linear equivariant is therefore $\alpha I + \beta M$ where $\alpha, \beta \in \mathbf{R}$ are our previous b_2, b_1 respectively. The eigenvalues and eigenvectors are:

$$u_0 = (1,1,\ldots,1)^T \quad \text{eigenvalue } \alpha + N\beta$$
$$u_1 = (1,-1,0,\ldots,0)^T \quad \text{eigenvalue } \alpha$$
$$u_2 = (0,\ldots,0,1,-1)^T \quad \text{eigenvalue } \alpha$$

There are two potential local bifurcation points: one where $\alpha + N\beta = 0$ and one where $\alpha = 0$. These are distinct when $\beta \neq 0$, a nondegeneracy condition that prevents the occurrence of a mode interaction point and is generically valid. If $\alpha + N\beta = 0$ then the kernel is V_0, whereas if $\alpha = 0$ then the kernel is V_1. We therefore choose α to be the bifurcation parameter, and arrange for the first bifurcation to have kernel V_1, with the trivial solution stable for $\alpha < 0$. This is the case provided $\beta < 0$. By scaling x we may assume $\beta = -1$, leading to a vector field of the form

$$F(x) = \begin{bmatrix} \lambda x_1 - (x_1 + \cdots + x_N) \\ \lambda x_2 - (x_1 + \cdots + x_N) \\ \cdots \\ \lambda x_N - (x_1 + \cdots + x_N) \end{bmatrix} + \text{h.o.t.}$$

This is the form required for simulating the full system on $\mathbf{R}^N$. However, it is much simpler to analyse the centre manifold reduction onto V_1 (or if all we are interested in is the equilibrium states, the Liapunov-Schmidt reduction) as we now describe.

Centre Manifold Reduction

As already indicated, the general cubic truncation of an $\mathbf{S}_N$-equivariant ODE involves 13 parameters, which is too complicated for detailed analysis. Here we describe work of Elmhirst [27] on the dynamics of the centre manifold reduction of such a cubic truncation to the space V_1, see also Cohen and Stewart [17]. Carr [13] is a useful source for information on the generalities of centre manifold reduction.

In Appendix 2 we show that the general cubic centre manifold reduced equations on V_1 take the form

$$
\begin{aligned}
dx_i/dt \quad = \quad & \lambda x_i + (Nx_i^2 - \Sigma_2) + C(Nx_i^3 - \Sigma_3) \\
& + D(Nx_i(x_1^2 + \cdots + x_{i-1}^2 + x_{i+1}^2 + \cdots + x_N^2) - \Sigma_{12})
\end{aligned}
\tag{9}
$$

where $1 \leq i \leq N$ and

$$
\begin{aligned}
\Sigma_2 &= x_1^2 + \cdots + x_N^2 \\
\Sigma_3 &= x_1^3 + \cdots + x_N^3 \\
\Sigma_{12} &= \sum_{i \neq j} x_i^2 x_j
\end{aligned}
$$

This equation has been analysed in considerable detail by Elmhirst [27], and we state some of his results without proof. We study the primary branches of equilibria, on the spaces $W_p = \mathrm{Fix}(\mathbf{S}_p \times \mathbf{S}_q)$ where $p + q = N$ and $1 \leq p \leq N/2$. These are the branches guaranteed by the Equivariant Branching Lemma. We also study the stabilities of these branches. We coordinatise W_p by $\alpha \in \mathbf{R}$, corresponding to the point

$$
\alpha(\underbrace{q, \ldots, q}_{p}; \underbrace{-p, \ldots, -p}_{q})
$$

Define

$$
\begin{aligned}
n &= N(N - 2p) \\
c &= C(N^2 - 3Np + 3p^2) \\
d &= D(N^2 - N(N + 3)p + (N + 3)p^2)
\end{aligned}
$$

Then the branching equation on W_p turns out to be

$$
\lambda = -\alpha n - \alpha^2 N(c - d)
$$

which is a parabola passing through the origin. Together with the trivial solution, the bifurcation diagram on W_p looks like Figure 2. Let α_0 be the α-coordinate of the other point at which the parabola crosses the α-axis, and let (λ_c, α_c) be the coordinates of the vertex of the parabola. Then

$$\alpha_0 = \frac{n}{N(d-c)}$$

$$\alpha_c = \frac{n}{2N(d-c)}$$

$$\lambda_0 = \frac{n^2}{4N(c-d)}$$

Stability

In order to compute the eigenvalues along the primary branches we use the isotypic components of the action of the isotropy subgroup to block-diagonalise the Jacobian. We therefore compute the isotypic components. The isotypic decomposition of V_1 for the action of $\Sigma = \mathbf{S}_p \times \mathbf{S}_q$ is

$$X = Y_0 \oplus Y_1 \oplus Y_2 \tag{10}$$

where

$$Y_0 = \{(\underbrace{qu, \ldots, qu}_{p}; \underbrace{-pu, \ldots, -pu}_{q}) : u \in \mathbf{R}\}$$

$$Y_1 = \{(\underbrace{x_1, \ldots, x_p}_{p}; \underbrace{0, \ldots, 0}_{q}) : x_1 + \cdots + x_p = 0\}$$

$$Y_2 = \{(\underbrace{0, \ldots, 0}_{p}; \underbrace{x_{p+1}, \ldots, x_N}_{q}) : x_{p+1} + \cdots + x_N = 0\}$$

Note that when $p = 1$ the component Y_1 should be omitted. Note also that $Y_0 = \mathrm{Fix}(\Sigma)$.

The action of Σ is absolutely irreducible on each isotypic component, and trivial on Y_0. We define basis elements for the Y_j as follows (where for simplicity we henceforth omit underbraces):

$$Y_0 \;:\; \xi = (q, \ldots, q; -p, \ldots, -p)^T$$

$$Y_1 \;:\; \eta_1 = (1, -1, 0, \ldots, 0; 0, \ldots, 0)^T$$

$$\eta_2 = (0, 1, -1, 0, \ldots, 0; 0, \ldots, 0)^T$$

$$\cdots$$

$$\eta_{p-1} = (0, \ldots, 0, 1, -1; 0, \ldots, 0)^T$$

$$Y_2 \;:\; \zeta_1 = (0, \ldots, 0; 1, -1, 0, \ldots, 0)^T$$

$$\zeta_2 = (0, \ldots, 0; 0, 1, -1, 0, \ldots, 0)^T$$

$$\cdots$$

$$\zeta_{q-1} = (0, \ldots, 0; 0, \ldots, 0, 1, -1)^T$$

Thus there are (at most) three distinct eigenvalues, one for each Y_j. We may therefore compute these eigenvalues by applying $DG|_{x_\alpha}$ to ξ, η_1, ζ_1, from which we deduce that

$$
\begin{aligned}
\mu_0 &= (G_{11} + \cdots + G_{1p}) - \tfrac{p}{q}(G_{1.p+1} + \cdots + G_{1N}) \\
\mu_1 &= G_{11} - G_{12} \\
\mu_2 &= G_{p+1.p+1} - G_{p+1.p+2}
\end{aligned}
$$

Putting all this information together we find that the eigenvalues of the linearisation along the primary branch in Fix(Σ) are:

$$
\begin{aligned}
\mu_0 &= \alpha n + 2\alpha^2 N(c - d) \\
\mu_1 &= \alpha N^2 + \alpha^2 N^2(2N - 3p)(C - D) \\
\mu_2 &= -\alpha N^2 + \alpha^2 N^2(3p - N)(C - D)
\end{aligned}
\tag{11}
$$

with multiplicities $1, p-1, q-1$ respectively. Note than when $N = 3$ we have $p = 1$ so μ_1 does not occur.

Observe that near the origin, where the linear term in α dominates, the eigenvalues μ_1 and μ_2 have opposite signs. Therefore the bifurcating branches are always unstable near the origin. This is why we must work with (at least) the cubic truncation.

We summarise a few of the main consequences of these computations. Ignoring degenerate cases when $C = D$ or $c = d$, each eigenvalue changes sign twice along the branch: once at the origin and once somewhere else. Let the non zero value of α at which this sign change occurs be β_j for μ_j. Then

$$
\begin{aligned}
\beta_0 &= \frac{n}{2N(d - c)} = \alpha_c \\
\beta_1 &= \frac{1}{(2N - 3p)(D - C)} \\
\beta_2 &= \frac{1}{(3p - N)(C - D)}
\end{aligned}
$$

Stability at Infinity

One interesting condition is that the branch should be stable for sufficiently large α ('stable at infinity'). This ensures that the speciation event persists for all sufficiently large λ – it is 'permanent'. It is easy to derive a necessary and sufficient condition for stability at infinity, as follows. The sign of μ_j near infinity is dominated by the coefficient of α^2, and we want all $\mu_j < 0$. Therefore we require

$$
\begin{aligned}
c - d &< 0 \\
(2N - 3p)(C - D) &< 0 \\
(3p - N)(C - D) &< 0
\end{aligned}
$$

where we are assuming $C \neq D$ and $p \neq N/3$. If $C > D$ then $p > 2N/3$, contrary to $p \leq N/2$, so we need

$$C < D \tag{12}$$

This being so, we deduce that $p > N/3$, and these conditions ensure that $\mu_1 < 0$ and $\mu_2 < 0$. It remains to consider μ_0. This is negative provided $c < d$, or equivalently

$$C < D \left(1 - \frac{Np(N-p)}{N^2 - 3Np + 3p^2} \right) \tag{13}$$

Both $Np(N-p)$ and $N^2 - 3Np + 3p^2$ are positive, and it is easy to prove that whenever $1 \leq p \leq N/2$ we have $Np(N-p) > N^2 - 3Np + 3p^2$, implying that $(1 - \frac{Np(N-p)}{N^2-3Np+3p^2}) < 0$. Thus if $D > 0$ inequality (13) implies (12), but if $D < 0$ then inequality (12) implies (13).

We deduce that necessary and sufficient conditions for a branch to be stable at infinity are:

1. If $D > 0$ then $C < D(1 - \frac{Np(N-p)}{N^2-3Np+3p^2})$ and $p > N/3$
2. If $D < 0$ then $C < D$ and $p > N/3$

The most interesting condition here is that in either case $p > N/3$, which goes a long way towards verifying a conjecture of Cohen and Stewart [17] to the effect that the bifurcating species must contain more than one third of the total number of PODs, and (since $p+q = N$) less than two thirds of the total number of PODs. That is, in this model, on the extra assumption of stability at infinity, the 'founder populations' are *large*. This is very different from the common assumption of a small founder population in the allopatric mechanism.

Of course, stability at infinity is an artificial condition, because it allows λ to increase without limit. Stable speciation can occur for bounded λ when $p \leq N/3$; indeed it can occur for $p = 1$. Nonetheless, there is a general tendency for speciation to 'prefer' large values of p in this model.

6. Simulations

We describe the results of some numerical simulations of (8). In these simulations we begin with a random initial condition near 0. At each time-step we 'ramp' the bifurcation parameter. That is, we increment λ by some fixed small amount and use the previous value of x as the initial condition for a single time-step in a numerical algorithm for solving the ODE (8). Here we have used the Euler method because of its simplicity, although a Runge-Kutta algorithm would be more usual. We add a small amount of random noise to each component of x: see Appendix 3. A low level of noise aids the numerics by preventing variables from becoming 'trapped' very close to zero. Finally, we found that if (8) is integrated without further precautions numerical solutions can diverge from V_1 and blow up. We therefore project x back onto V_1 at each integration step by subtracting the mean of the $x(j)$ from each component.

Figure 7 shows a typical speciation event, occurring for the parameter values shown in the caption. Here the number of PODs is $N = 25$, and the initial single-species state splits into a state with $p = 8, q = 17$. The curves are slightly irregular because we have included a moderate amount of noise. The bifurcation is relatively rapid, but may appear not to deserve the 'jump' description. There are two reasons for this. First, if the bifurcation parameter is ramped sufficiently slowly, the jump can be made as close to the vertical as desired. Second, it is well known that ramping can lead to the phenomenon of 'tunnelling through the bifurcation'. Prior to bifurcation, $x = 0$ is a stable equilibrium so solutions converge rapidly towards it. Immediately after bifurcation $x = 0$ is only weakly unstable, so it takes some time before x starts to diverge from 0. We have not taken steps to eliminate this 'slow passage' effect because we feel that ramping the bifurcation parameter is very much in the spirit of real evolutionary dynamics, where each generation forms the 'initial conditions' for the next, and the environment slowly changes. There is an extensive literature on slow passage effects: references include Baer *et al.* [3], Candelpergher *et al.* [12], Diener and Diener [24], Hayes [40], Neihstadt [62, 63], and Su [76, 77]. Of these, references [3, 62, 63] point out that noise destroys slow passage effects.

Since $8 < 25/3$ we know from general theory that when λ becomes sufficiently large, the solution depicted must become unstable. Nevertheless, we see that it can remain stable for a broad range of λ values.

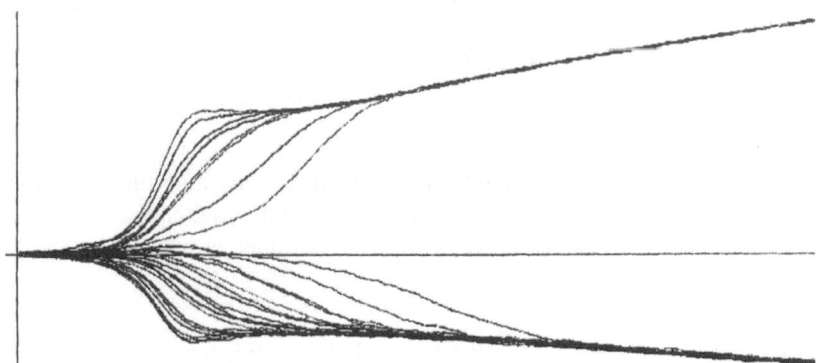

FIGURE 7. Numerical simulation of (8). Here λ runs horizontally from -50 to 1000, and each component of x is plotted on the same vertical axis. Parameters are $N = 25, C = -1, D = -.2$.

In Figure 8 we have drawn the parabolas formed by the primary branches on each fixed-point space W_p, where for clarity we shown only $p \leq 6$. A thick line shows where all eigenvalues are negative, that is, the primary solution on that branch is stable. The vertical scale for the parabolas bears no meaningful relation to that for the x_j. In this simulation we have set the noise to a very low level, and the curves are smooth.

Immediately after bifurcation *no* primary branches are stable. Before this causes a problem, however, the bifurcation parameter ramps to a value at which the $p = 1$ branch becomes stable, and we see that the solution does indeed converge to a state with $p = 1$. Subsequently, as λ increases further, the states $p = 2$ and $p = 3$ become stable. Next, $p = 1$ loses stability. Then state $p = 4$ becomes stable, then $p = 5$, and so on (off the picture).

The numerical solution remains in the $p = 1$ state until shortly after that state loses stability. It then makes a rapid transition to a $p = 3$ state, which is stable at that value of λ. (So are $p = 2, 4$: presumably different random perturbations might have taken the system to one of those states; or there may be some general constraints on such transitions of which we are unaware. In other simulations at these parameter values, with different initial seeds for the random number generator, we have seen a transition to a $p = 2$ state.) Again, there is a slight delay resulting from tunnelling through the bifurcation.

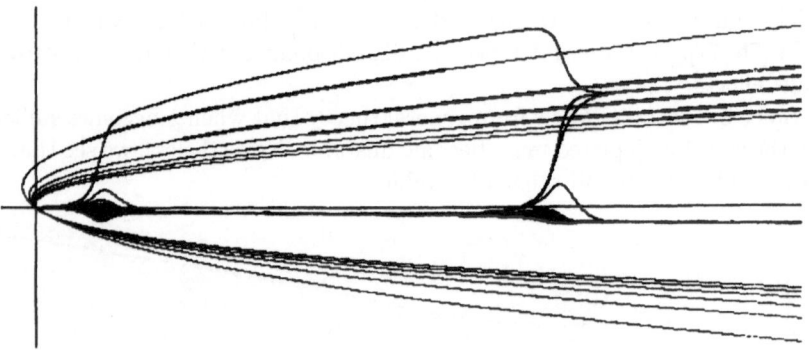

FIGURE 8. Numerical simulation of (8) illustrating a secondary bifurcation. Here λ runs horizontally from -50 to 1000, and each component of x is plotted on the same vertical axis. Parameters are $N = 25, C = -.31, D = -.2$.

In Figure 9 we have removed the parabolas, using horizontal black lines to show the intervals in which all eigenvalues are negative. Here p is 1 for the lowest such line and increases by 1 for each upward step. Again we set the noise to a very low level. This time $N = 11$.

Again, immediately after bifurcation no primary branches are stable. The bifurcation parameter ramps to a value at which the $p = 1$ branch becomes stable, and the solution converges to a state with $p = 1$. When that state loses stability, it switches to a state with $p = 2$. When *that* state loses stability, it switches to a state with $p = 3$. Since $3 < 11/3$ we can predict that for larger λ, off the picture, the state would switch again, probably to $p = 4$ which is stable at infinity (but

$p = 5, 6$ are also possible). Again, we see a slight delay resulting from tunnelling through the bifurcation.

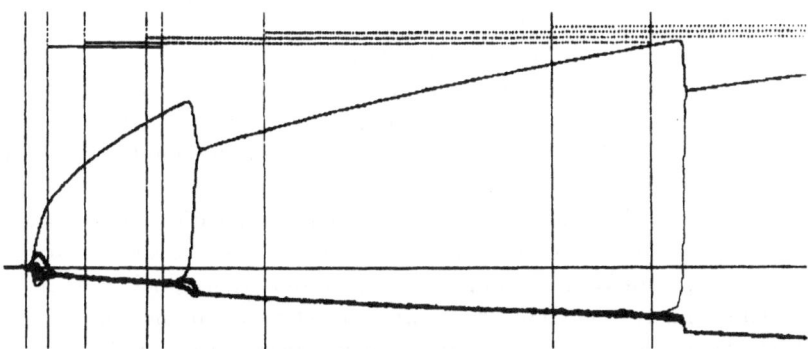

FIGURE 9. Numerical simulation of (8) showing two successive secondary bifurcations. Here λ runs horizontally from -50 to 1000, and each component of x is plotted on the same vertical axis. Parameters are $N = 11, C = -10, D = -9$.

Figure 10 repeats this scenario but with higher noise. The noise broadens the curves into bands. It also causes 'premature bifurcation' in which the solution moves away from an equilibrium before it loses stability. What is happening here is that at least one eigenvalue, though still negative, is getting close to zero, and the size of the basin of attraction is becoming very small. Large enough noise can kick the system out of that basin into some other, more robust basin.

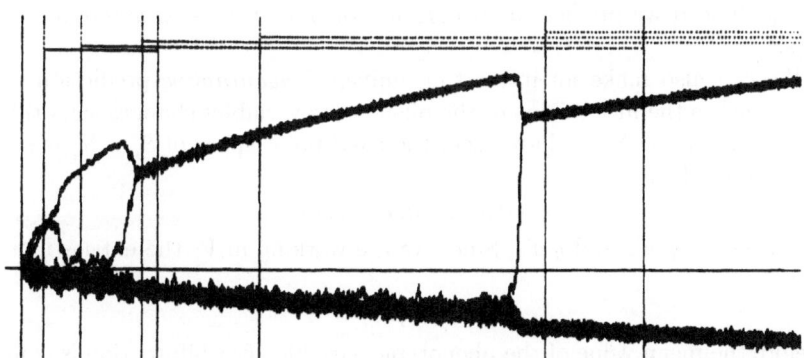

FIGURE 10. Numerical simulation of (8) at same parameter values as Figure 9, but with higher noise. The noise causes premature secondary bifurcations.

7. Abstract Analysis

We now show that the general methods of equivariant bifurcation theory, outlined in §4, make it possible to analyse the equilibria of (8) by exploiting the symmetry explicitly.

We can find symmetry-breaking equilibria of a symmetric dynamical system by applying the Equivariant Branching Lemma. The number of species present in a solution x of (9), or any other $\mathbf{S}_N$-equivariant system of ODEs, is given by the number of distinct entries in x, and we can find the number of species by computing the isotropy subgroup Σ_x of x.

We have seen that bifurcation occurs when the linearisation $DF|_0$ becomes singular. Generically the kernel of the linearisation is one of the two irreducible components V_0, V_1 of $\mathbf{R}^N$. Both of these are absolutely irreducible. If the kernel is V_0, then symmetry does not break: we just get a new branch of $\mathbf{S}_N$-invariant equilibria and the population remains a single species, though the species as a whole experiences a continuous change in phenotype. The case where the kernel is V_1 leads to symmetry-breaking and bifurcation to more than one species. We therefore restrict our analysis to this case.

By the Equivariant Branching Lemma, if there is a steady-state bifurcation with kernel V_1 then there exist branches of solutions for all axial isotropy subgroups. The axial subgroups of $\mathbf{S}_N$ in this representation are, up to conjugacy, those of the form $\mathbf{S}_p \times \mathbf{S}_q$ where $p + q = N$ and $1 \leq p \leq [N/2]$. So there exist branches of solutions with these isotropy subgroups. Such solutions lie in fixed-point spaces of the form $(u, \ldots, u; v, \ldots, v)$, with exactly *two* distinct values u and v for phenotypic variables. Such solution branches therefore correspond to a split of the population of N identical PODs into two distinct species consisting of p and q PODs respectively. One species has the phenotype u and the other species has the phenotype v . So in this example speciation is a consequence of symmetry-breaking. Indeed we predict the occurrence of bifurcations to *dimorphism* – two species.

We can also make an interesting universal *quantitative* prediction: on the above branches the mean value of the phenotypic variables changes smoothly during the bifurcation. As we have seen, the fixed-point space of $\mathbf{S}_p \times \mathbf{S}_q$ is spanned by all vectors of the form

$$(u, \ldots, u; v, \ldots, v)$$

where there are p u's and q v's. Since we are working in V_1 the entries must sum to zero:

$$pu + qv = 0 \tag{14}$$

Therefore the mean value of the phenotypic variable after bifurcation is computed as $(pu + qv)/N = 0$. Before bifurcation it is also 0, because we are looking for bifurcations from the trivial equilibrium and have normalised the phenotypic variables to be zero for the original single species. So here the mean remains constant throughout the bifurcation. However, we are working with the centre manifold reduction, which involves a nonlinear change of variables. Therefore the mean varies

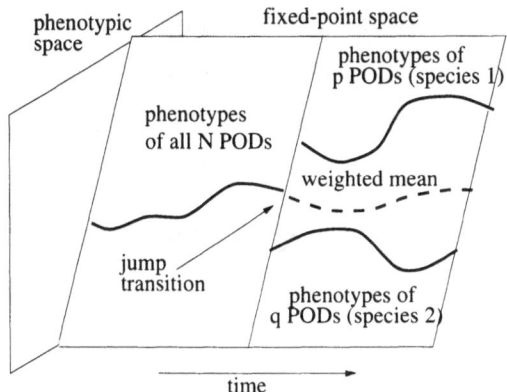

FIGURE 11. Generic $\mathbf{S}_N$-equivariant steady-state bifurcation: jump transition from one species to two, with smoothly varying mean.

smoothly in the original phenotypic variables, and is thus approximately constant, as illustrated schematically in Figure 11.

8. BirdSym

Our analysis so far has been abstract: not based on any explicit interpretation of the phenotypic variables. The results obtained are thus *model-independent*: they are valid whatever that intepretation may be. However, we can make further progress by specialising the model to phenotypic variables with a specific biological interpretation.

Elmhirst [28] has investigated a simulation, BirdSym, loosely based on Darwin's finches, whose variables have specific biological interpretations. It can be reduced to an $\mathbf{S}_N$-equivariant dynamical system. We summarise the construction of the BirdSym model and describe sample results from simulations.

We assume a population of N PODs, composed of birds. POD i has phenotype $z_i = (x_i, y_i)$ where x_i is beak length and y_i is beak variance. The environment provides a single resource (which we refer to as *seeds* but think of as exploitable energy). The resource continuum is discretised and the resource mean is a_1, its variance is a_2, and the global abundance is a_3. The amount of energy available in size l seeds, in appropriate units, is

$$R(l) = \frac{a_3 e^{-(l-a_1)^2/2a_2}}{\sqrt{2\pi a_2}} \delta l$$

The efficiency with which birds in POD i can eat seeds of size l is assumed to be

$$f_i(l) = \frac{e^{-(l-x_i)^2/2y_i}}{\sqrt{2\pi y_i}}$$

The total amount of energy that bird i accumulates is therefore

$$E_i = \sum_l \frac{R(l) f_i(l)}{F_T(l)}$$

where

$$F_T(l_j) = \sum_{i=1}^{N} f_i(l)$$

On the assumption that PODs move up energy gradients, the dynamics can be described by a system of ODEs

$$dx_i/dt = g_i(x_1, \ldots, x_N; a)$$

where

$$g_i(x; a) = C\left(\frac{-N x_i e^{-x_i^2/2\lambda}}{\lambda\sqrt{\lambda}} - \sum_{j=1}^{N}(x_j - x_i)e^{-(x_i-x_j)^2/2} \right) \tag{15}$$

and the parameter $C > 0$ is a measure of how fast the pods can adapt. This ODE is determined by an $\mathbf{S}_N$-equivariant smooth vector field, but is not polynomial.

A typical numerical solution of (15) is shown in Figure 12. Note the repeated bifurcations as the bifurcation parameter λ increases, which is typical of BirdSym. We return to this phenomenon in §10 in connection with the periwinkle *Littorina saxatilis* and various subspecies of grasshopper.

Elmhirst relates BirdSym to (8) by approximating the exponentials by their Taylor series. He then uses the analysis of (8) to study the bifurcation behaviour, and compares the analytic results with numerical ones. The correspondence is excellent. He also derives further features of BirdSym, showing that for the full equations (15) there exist parameter values at which all primary branches are stable. He also, intriguingly, finds that there exist parameter values at which solutions oscillate periodically: an example is shown in Figure 13.

The main conclusion of Elmhirst [28] is that sympatric speciation occurs because the common 'generalist' strategy used by the single-species population to exploit its environment becomes dynamically unstable, and is replaced (because of symmetry) by a combination of two 'specialist' strategies. That is, the *task space* – the space of different ways to exploit the environment – becomes partitioned between different subsets of the population.

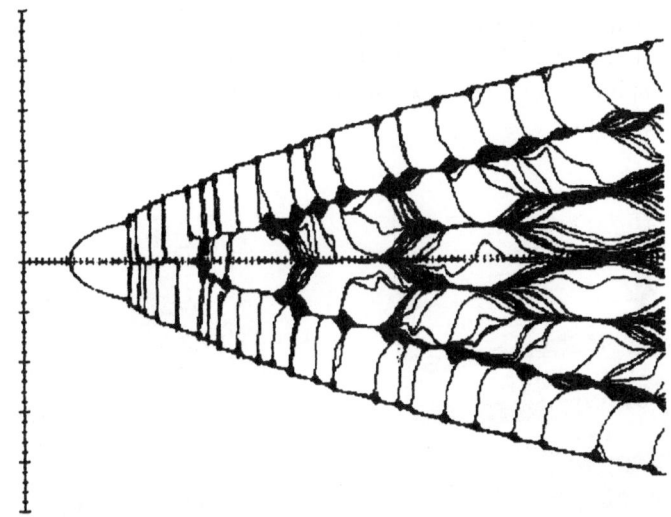

FIGURE 12. Simulation of BirdSym with $N = 120$.

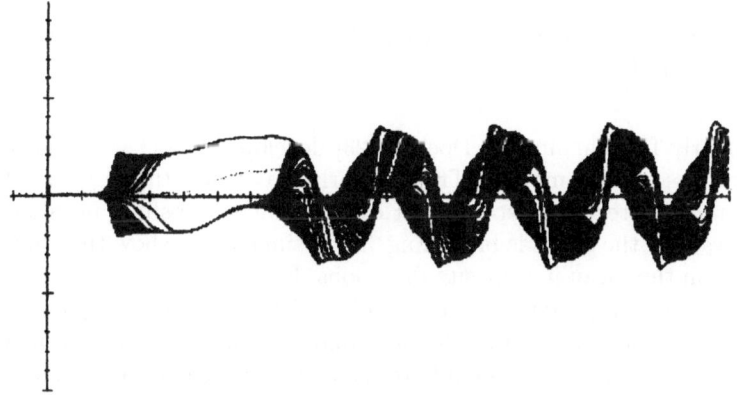

FIGURE 13. Periodic oscillations in a BirdSym simulation.

9. Darwin's Finches and Other Field Observations

Is there any evidence in favour of the symmetry-breaking model of sympatric speciation? Some studies reported in the existing literature are reasonably consistent with its predictions, especially the 'constant mean' prediction. These include various biologically motivated models, such as Higashi *et al.* [42], Kondrashov and Kondrashov [53], Dieckmann and Doebeli [23], and Tregenza and Butlin [79].

In particular, Higashi *et al.* [42] argue that sympatric speciation can be accomplished through sexual selection without disruptive natural selection. Their model comprises a population of sexually reproducing organisms, from which N

males and N females are chosen at random at each step. They study how the probability distributions of female preference and male phenotype coevolve, finding that each splits into two groups, diverging from the original mean in opposite directions, Figure 14.

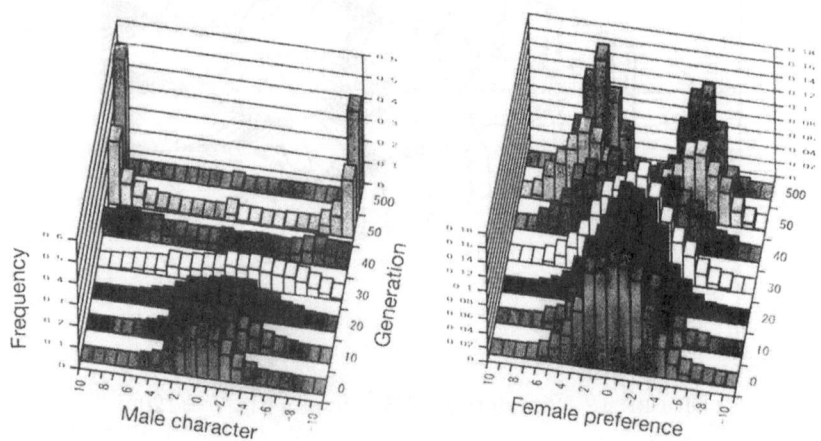

FIGURE 14. Sympatric speciation through sexual selection, after Higashi *et al.* [42].

Similarly Dieckmann and Doebeli [23] develop a model with multilocus genetics and assortative mating. They remark that when the original phenotype (ecological character) loses stability, it can then be successfully invaded by other phenotypes, and the result is branching. Their simulations show that the branches split off from the mean in opposite directions, Figure 15.

The process of speciation cannot be observed directly on geological timescales, for obvious reasons, although a certain amount can be inferred from the fossil record. It is also sometimes possible to make direct observations of speciation 'in miniature' in such organisms as Darwin's finches, African lake cichlids, sticklebacks, and fruit flies. More commonly, we can observe what appears to be the end result of a speciation transition, with two closely related species or subspecies coexisting in a given environment (which may be a 'hybrid zone' where the two species's normal habitats are adjacent). This is the 'allopatric' context. Having observed the phenotypes that occur in the allopatric context, we can compare them with the corresponding phenotypes for the two species when only one of them exists (ideally in the same environment as for the allopatric situation). This is the 'sympatric' context.

With this interpretation, the predictions of the model, especially that the mean should be the same in either allopatric or sympatric populations, are also consistent with some field observations, some of which we now list. Beauchamp and Ulyett [10] report a preference for temperature ranges in the flatworms *Planaria*

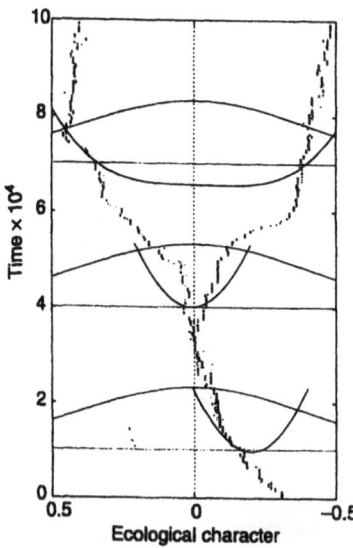

FIGURE 15. Sympatric speciation through assortative mating, after Dieckmann and Doebeli [23].

montenegrina and *Planaria gonocephala* which occupy similar ranges in the allopatric case but non-overlapping ranges in the sympatric case. Huey *et al.* [46, 44], Huey and Pianka [45] describe phenotypic differences (vent-snout length) in subterranean skinks *Typhlosaurus lineatus* and *Typhlosaurus gariepensis*. Bantock and Bayley [4] and Bantock *et al.* [5] discuss shell sizes in the snails *Cepea nemoralis* and *Cepea hortensis* resulting from selective predation by birds. Fenchel [29, 30] reported differences in allopatric and sympatric populations of the mud-snails *Hydrobia ulvae* and *Hydrobia ventrosa* (but see Barnes [7, 8, 9] for caveats).

Polymorphism in Darwin's Finches

A celebrated instance of polymorphism is the changes in beak size that occur among various species of Darwin's finches in the Galápagos Islands, and we describe this as a typical example. The evolution of the different finch species in the Galápagos Islands is thought to have occurred around five million years ago, and so cannot be observed (although small-scale evolution remains rapid enough that significant phenotypic changes can be observed from one year to the next). However, as just explained, we can observe a surrogate for actual evolution: differences in the phenotype of a given species in allopatric and sympatric populations. The transition in phenotype from sympatric populations to allopatric ones should be just like the bifurcations in the speciation model: in particular, we expect to see approximately the same mean in either situation.

Lack [54] showed that mean beak size varies between islands, and Grant *et al.* [38] showed that the standard deviation of beak size is similarly variable. Here

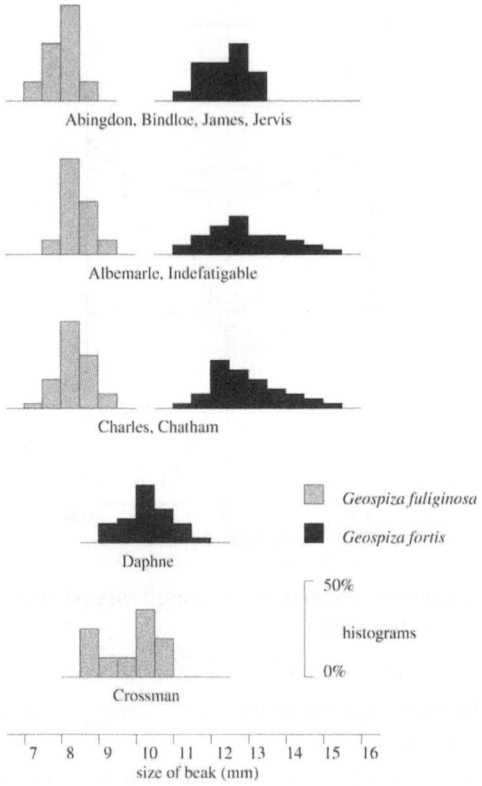

FIGURE 16. Beak sizes in allopatric and sympatric populations of *Geospiza* in the Galápagos Islands. After Lack [54].

we focus upon the two species *G. fortis* and *G. fuliginosa*, which occur in both sympatric and allopatric populations. *G. fortis* is allopatric on Daphne, and *G. fuliginosa* is allopatric on Crossman; the two species are sympatric on a number of islands which Lack placed in three groups for data analysis: Abingdon, Bindloe, James, Jervis; Albemarle, Indefatigable; and Charles, Chatham. Figure 16, adapted from Lack [54], shows the differences in beak size between these species on the cited groups of islands. The mean beak sizes of both *G. fortis* and *G. fuliginosa* are approximately 10mm in allopatric populations. In all three (groups of) sympatric populations, the mean for *G. fortis* is about 12mm, while that for *G. fuliginosa* is about 8mm. These figures are consistent with the 'constant mean' prediction. Simply from the biology, constant mean would *not* be expected; therefore we find confidence in the mathematical model from its prediction of a biologically 'unnecessary' constraint.

10. Biological Interpretation

The usual story told to explain changes in an organism's phenotype is that genetic mutations change its genotype, and this in turn modifies the phenotype. Phenotypic change is therefore seen as the result of a gradual accumulation of mutations, with the principal agent of change being the gene.

Sometimes this is an appropriate picture. For example, Ridley [66] discusses the lacewing species *Chrysopa carnea* and *C. downesi*. *Carnea* is light green (in summer, brown in autumn), *downesi* dark green. They have different habitats and different breeding times: no natural hybrids occur. They can be made to interbreed by artificially varying the hours of 'daylight', so in this sense they are 'really' one species. It is generally thought that a colour polymorphism arose sympatrically via a single mutation, after which natural selection favoured habitats that made good camouflage – light-coloured grass, meadows and deciduous trees for *carnea*, dark coloured conifers for *downesi*. A subsequent mutation in a gene that changed the photoperiodic response then made the breeding times diverge.

Such cases may not be entirely representative, however. The rate at which a given DNA base-pair mutates is about 1 mutation every 10^9 years, Ridley [67]. In the human genome (3×10^9 base pairs) this amounts to 3 point mutations per year, so gradual accumulation must be *very* gradual. In the orthodox, mutational view, many mutations have no effect – they are 'neutral' – because of redundancy in the genetic code, because they are in unexpressed segments of the genome, or because they are lethal and promptly eliminated. The concept of neutral mutations was proposed by Kimura [52] to account for the enormous variety of alleles in natural populations, discovered by Lewontin and Hubby [55], and totally unexpected by classical population genetics. Superficial expectations that alleles would be neutral in some cases (for example if they code for the same amino acid) have been shown to be mistaken (because different transfer RNAs for different codons are available in different quantities in different cells). It was also believed that natural selection would be unable to discriminate between alleles with different numbers of repeats in 'satellite' sequences, but we now know that many human diseases – for example 'fragile-X' – have this basis. The 'neutral allele' hypothesis has largely failed. Allele differences do make a difference: not singly, but in combination.

Recombination, not Mutation

A far more potent force for phenotypic change, in sexual species but also in others (for example bacteria) is recombination. The offspring inherit two alleles at each genetic locus: one from the father, one from the mother. They are *homozygous* at a locus if these alleles are the same, and *heterozygous* otherwise. In Fisher's population genetics, it was expected that organisms would be homozygous at almost all loci. Lewontin and Hubby [55] showed that, on the contrary, a typical individual is heterozygous at about 10% of loci, while over 30% of loci are heterozygous across the entire population. This implies that genetic variability between and among individuals is vastly greater than Fisher believed. As a result, recombination of alleles from parents creates far more genetic diversity than mutations, and it does so without fail at every reproductive step. Each generation shuffles the alleles of the previous generation: cryptically inherent in the gene-pool of the species is a huge range of potential phenotypes. Changes in environment or selective pressure lead to a rapid change in the phenotypes observed *in surviving adults*. Many species produce far more offspring than can survive to adulthood. For example a female starling lays about 16 eggs during her lifetime, of which on average 14 die without breeding; a female frog lays some 10,000 eggs, of which on average 9,998 die without breeding; a female cod lays 40,000,000 eggs, of which on average 39,999,998 die without breeding. Such environmentally-dependent elimination of large numbers of genotypes from the next generation accelerates the process of phenotypic change. This post-1970s view of the genetic basis of evolutionary change – more Wallace than Darwin – is much more congenial to our mathematical model.

Non-Genetic Influences

There are also numerous non-genetic or quasi-genetic influences on phenotype. For example:

- Maternal-effect genes: the early development of the egg is controlled by certain genes of the mother, not of the developing organism.
- Privilege: the parents can provide non-genetic assistance to the developing offspring such as yolk, milk, a nest, food.
- Genetic assimilation and related effects, such as the Baldwin effect: environmental stress can reveal the effects of cryptic genetics, giving rise to what superficially looks like Lamarckian evolution.
- Molecular 'epigenetics': the genome can be labelled (methylation) to activate or suppress particular genes; many other molecular tricks of this kind are now known. Again the effect often looks superficially Lamarckian. See Wolffe and Matzke [82], Francis *et al.* [32].
- Prions: it is now known that the yeast genome specifies some proteins in the wrong configuration, and these are converted to the correct form by prions; the prions are inherited directly rather than being constructed afresh using a protein-coding sequence in parental DNA.
- The chaperonin HP70 constrains mutated proteins into their unmutated shape. Only when 'stress' uses HP70 for other functions are the mutations expressed.

All this makes classical Fisherian genetics highly implausible, and poorly suited to discussing speciation. The linearity of classical genetics, in which a collection of organisms is replaced by a mean-field gene-pool and fitness is a linear combination of proportions of alleles, adds to the implausibility. Indeed, we will argue that when sympatric speciation of the kind implied by our model occurs, it will be invisible to classical genetics.

Mean-Field Genetics

The short story is that Fisherian genetics is a mean-field theory, and in our model speciation is invisible to the mean field (mean values of phenotypes do not undergo noticeable changes).

As an aside: another way to say this is that σ_1 is not a symmetry detective for the breaking of symmetry from $\mathbf{S}_N$ to $\mathbf{S}_p \times \mathbf{S}_q$, in the sense of Barany *et al.* [6]. In contrast, the *deviation* from the mean 'sees' the speciation event, but cannot distinguish the number of new species. Appropriate statistical measures (and/or symmetry detectives) for speciation events must involve the invariants σ_k, the 'higher moments' of the data, and in carefully chosen combinations. There is scope here for some useful research.

We explain this point in more detail. In the model (8) we have so far interpreted the x_i as phenotypic variables, but this interpretation is not mandatory. Suppose that some of the x_i are genetic variables, representing Fisher-style proportions of alleles (within the organisms of a POD). Suppose that certain linear combinations of these allele proportions determine certain continuous characters y_j. Then, when speciation occurs, the means of the genotypic x_i remain constant, and the y_j remain constant as well. That is, the continuous characters to which Fisherian assumptions apply cannot change in any detectable way, even though speciation occurs as a jump bifurcation in this model.

Of course, the model can be rejected. But it is still disturbing to realise that a linear mean-field theory is inherently insensitive to a reasonable mechanism for speciation, and that the reason for this insensitivity is reliance on means as primary observables.

Does this, perhaps, mean that sympatric speciation according to our model falls apart because it is incompatible with reasonable genetics? We believe not – and the key, again, is recombination, coupled with the fact that continuous characters are polygenic (and in a nonlinear, non-trivial way). We should also remember (as many Fisher-style models do not) the central role of *selection*. Selection eliminates organisms, not alleles. Selection can mitigate against particular *combinations* of alleles without removing those alleles from the population, and even *without changing their proportions*. Mean field theories are insensitive to the distribution patterns of their ingredients. Bean-counting tells you how many beans there are, but not how they are arranged or associated. A cake could consist of a layer of flour, a lump of butter, a heap of sugar, and a pool of egg, and a mean-field observer would be none the wiser.

Hybrid Zones

An instructive example occurs in a hybrid zone and involves grasshoppers, see Butlin and Hewitt [11] and Neems and Butlin [61]. It involves two (sub)species of grasshopper, *Chorthippus parallelus parallelus* and *Chorthippus parallelus erythropus*, which we henceforth call X and Y.

Chorthippus parallelus is a common European grasshopper whose habitat is moist meadows. The subspecies $X = C.\ p.\ parallelus$ occurs throughout most of

Europe, but is replaced by $Y = C.\,p.erythropus$ in the Iberian peninsula. The subspecies meet and hybridise in the Pyrenean mountains. In this 'hybrid zone' there is interbreeding, and the hybrid offspring are viable in the sense that they survive beyond the larval stage. However, they seldom survive to adulthood: instead, they die when they reach about a third of normal adult body-length. Genetically, the primary differences between the two species occur at about 20 loci, at each of which alleles can be of two types: call these A and B. In species X, all 20 alleles are of type A; in Y, they are all of type B. The hybrids have type A alleles at some loci and type B alleles at the other loci.

Fisher-style genetics can obviously distinguish one species from the other. However, it cannot distinguish a random mixture of X and Y, occupying territory in and around the hybrid zone, from the hybrids themselves. The overall proportions of alleles will be essentially the same in the mixed and the hybrid populations. The difference between the species lies in which *complexes* of alleles occur in breeding adults. In every generation, cross-breeding between X and Y re-creates the same overall gene-pool of all possible combinations of As and Bs. In every generation, selection then prunes away all combinations other than all As or all Bs.

Note that As and Bs are not 'good' or 'bad' genes. Those of X, labelled A, are not good in Y; those of Y are not good in X. Laboratory populations probably could carry either, but in the field the *combinations* do not permit maturity in competition with X (all A) and Y (all B).

Agreed, differences of habitat *do* distinguish the species, and since fitness coefficients can be assumed habitat-dependent we cannot conclude that Fisherian genetics breaks down completely here. Nonethless, we see one serious limtation of mean-field genetics when it is applied independently to several loci.

An example of a similar phenomenon, due to Reilly *et al.* [71, 72, 73], concerns cholesterol production in humans, which involves alleles at three loci. Again, no individual allele is good or bad in its own right, but particular combinations of alleles are.

PODs in Field Observations

How can PODs be observed? A POD is a coarse-grained cluster of organisms, but generally speaking, we have not specified how to determine such clusters in a reasonable manner, other than to point to the concepts of a deme (Salthe [70]) and a lineage (Rollo [68]). The important feature of a division of the population into PODs is that this division should reasonably support an evolutionary dynamic. This condition requires there to be a genetic link between successive generations of organisms belonging to the same POD. This condition holds for both lineages and demes, but we wish to propose an alternative that is often observed in the field.

Field observers who become sufficiently familiar with populations of organisms in 'the same' species often notice that the populations splits into clusters of organisms with similar behavioural patterns. For example, this occurs for the periwinkle *Littorina saxatilis*. Although most members of this species have the same external appearance, analysis of their DNA shows that they fall into well-defined clusters: what appears to be one species is really a large complex of subspecies, all subtly different from each other. The same goes for grasshoppers. It would be reasonable to take such behavioural/genetic clusters as the field equivalent of the theoretical concept of a POD.

11. Speciation Events

How would a speciation event occur according to our model? In order to attempt to answer that question we must endow the model with additional baggage – interpretations of the variables and parameters. Any such attempt will be tentative at this early stage. Nonetheless, if we graft on to the model a role for genetics that is consistent with current thinking, we can suggest a plausible scenario.

There are three main features: genes, organisms, environment. Genes affect organisms via development; environment affects organisms via selection. (More properly: environment sets the arena in which competition is played out, and this determines the 'rules' of the competitive game.) Organisms affect genes via reproduction, and they affect environment by existing within it and exploiting it. See Figure 17.

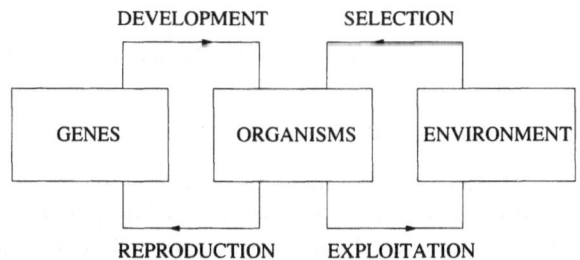

FIGURE 17. Schematic of evolutionary influences.

Note that the *organism* is where the whole picture comes together, not the gene.

The arrows in the diagram are seductive, and they conceal a number of difficulties. The players of the game change as time passes, so arrows can associate events that occur at different times. Moreover, the arrows are not mappings: each depends on an appropriate context, which includes all the others (and more). 'Environment' includes other organisms, *especially* those in the same species. Moreover, everything depends on local geography/habitat, which may vary in time as well as in space.

Figure 17 suggests that speciation is most reasonably viewed as a phenomenon occurring in a complex adaptive system: see Auyang [2], Kauffman [48, 49],

Mainzer [56]. This is a system of interacting entities (the organisms) whose states (phenotypes and genotypes) evolve according to some system of rules (selection). Speciation is an *emergent* property of the complex system, occurring on a higher level of description. By 'emergent' we mean that the derivation of the high-level property from the entities is computationally intractable, see Cohen and Stewart [15, 16].

A Finch Scenario

Consider a population of finches whose environment consists of seeds of various sizes. The role of the finches' genes will be to determine the sizes of their beaks. Since beak-size is a continuous character, it is presumably polygenic, involving not only the production of proteins for making beaks and getting them to the appropriate places in the developing embryo, but also such characters as the ability to develop larger beaks as a result of exercise (trying to crack bigger or harder seeds, or *more* smaller ones). In practice, these alleles will not affect *only* beak size – most genes affect most characters. But for illustrative purposes we will focus solely on their effect on beak size. We therefore postulate a complex of genetic loci $G_1, \ldots, G_k$ each with two alleles A_j or B_j. In general terms, A_j tends to produce short beaks, and B_j long ones. Again, in realistic circumstances the As and Bs will also have recombinational effects on other features of each bird. This contrasts strongly with the Fisherian assumption that any given allele has a fixed 'fitness' in any organism of a given species, Cook [19].

In such a simplified model, beak length is assumed to be some function of which alleles are present. To be specific, we could define ε_j to be -1 if locus G_j contains allele A_j, and $+1$ if locus G_j contains allele B_j. Suppose that

$$\text{beak length} = \alpha(\varepsilon_1 + \ldots + \varepsilon_k) + \beta$$

for constants $\alpha, \beta \in \mathbf{R}$. Here β is the mean beak size (assuming A_j, B_j are equally likely). A linear relation such as this will be at best an approximation, but if the functional relationship is smooth the approximation will be a reasonable one.

Assume that the sizes of available seeds are distributed unimodally about some medium-sized peak value, well suited to beaks of average size. Initially, the most efficient way for the finches to exploit the seeds is to specialise on medium-sized seeds. Most birds therefore have a mixture of alleles of types A and B, in roughly equal proportions (each A cancels one B and vice versa). *All* alleles A_j, B_j are abundant in the population and there is no systematic association between them in individual finches. (There are, however, global contraints on how these alleles associate: neither A nor B should predominate.) The finches are phenotypically indistinguishable and form a single species.

Next, suppose that the environment changes, in a manner that affects all finches uniformly (no 'geographic' variability). Suppose, furthermore, that this change lowers the survival value of specialising on medium-sized seeds. This could happen, for example, if the finch population began to outstrip its resources at that size of seed, or if the seed distribution broadened. (It is not necessary to assume that the seed distribution becomes bimodal, creating two 'niches' which the finch phenotypes follow: see Elmhirst [28] for mathematical justifications of these assertions. However, such bimodality might also lower the survival value of the medium-sized-seed strategy.)

Previously, it paid for all finches to have the same sized beaks as the others, but now that does not work. Individual finches survive better by adopting some other strategy – but their survival also depends on what strategies the others adopt. For example, if all finch phenotypes switch to larger beaks, or they all switch to smaller ones, the problem just becomes worse.

Of course, the finches cannot 'know' this (we are not talking species selection, kin selection, or any form of collective cooperation or knowledge, see below). Instead, random mating within the population produces individuals with more As than average, who eat smaller seeds, and with more Bs than average, who eat larger ones. In the changed conditions, *both* of these groups prosper; moreover, they do not compete with each other because they prefer seeds of different sizes. Therefore, *between them*, they can now out-compete the finches that stick to medium beak sizes.

Selection on individuals plus random recombination now favours a mixed strategy for the group: some finches with a predominance of As, others with a predominance of Bs. Others with a more balanced genome are eliminated.

When the two groups start to separate, gene-flow between them is initially unimpeded. So why do not the two groups merge? The answer is that the *same* selective pressures eliminate those hybrids that happen to have As and Bs in roughly equal numbers. In each new generation the full range of genotypes is re-created (all alleles remain present in the gene-pool); middle-of-the-road genomes are repeatedly weeded out by competition, as for the hybrid grasshoppers.

In the longer run, other factors come into play that can stabilise this split, the most obvious being assortative mating. We assume environmental uniformity in order to show that *even at this extreme* of homogeneity, speciation is possible and sometimes inevitable. Additional environmental inhomogeneities make speciation easier, not harder; they probably occur in a geographically patchy way. Patchy environment reinforces assortative mating: the commonest opportunities for mating involve partners who prefer the same kind of environment.

Our models add two important features, not intuitive in a verbal story. First, the initial separation is rapid, and the diverging phenotypes differ by a relatively large amount. For example, beaks sizes jump rapidly to new values, clearly distinguishable from old ones. Second, the split is hysteretic: once it has occurred, it does not necessarily reverse if the environment returns to its original state. (It *can* be reversed, perhaps temporarily, by bigger environmental changes, but not by small fluctuations.)

Genetically, then, what we observe is a sudden and robust reassortment of the genomes of surviving, breeding adults into two types: one with mostly As, the other mostly Bs. The overall proportions of single alleles do not change significantly.

The scenario just described is qualitatively similar to what happens in the grasshopper hybrid zone descibed earlier. Our model makes certain quantitative predictions, notably the constancy of the mean, which in principle could be tested against field data.

Species, Kin, Group, and Ecosystem Selection

Our model sheds some light on a long-running disagreement. The core Darwinian story is one of selection acting on individual organisms. There have been numerous controversies about the possibility of a form of selection acting on other 'units', such as species, herds, or family groups, see Ridley [67] chapter 12. Neo-Darwinism (Dawkins [20, 21]) claims to reduce all selection on organisms to selection for effects of genes, and therefore considers all debates about 'the unit of selection' to be answered by one word: 'gene'.

An excellent history of such controversies, with a substantial list of references, can be found in Hecht and Hoffman [41]. At the time when such issues were being debated, most analyses were couched in terms of the linear models of classical population genetics. In such models, selective advantages that operate on a group can normally be exploited by maverick individuals, who 'cheat' by not cooperating with the group behaviour, see Ridley [67] p.326. This effect was generally deemed to have demolished any chance of selection on the group level – except perhaps when the group comprises closely related individuals with many shared alleles – because it eliminates any presumed 'evolutionary advantage' in the group behaviour. This approach has led to an extensive literature on the 'bean-counting' view of altruism: for example, risking your own life by jumping into a river to save your brother offers evolutionary advantages provided the chance of not drowning is greater than 50%, because he 'shares half your genes'. (See Ridley [67] p.321 for the general criterion for such choices.) On the other hand, doing the same for a total stranger is never worthwhile.

The whole game goes out of the window, however, as soon as the linear models are rejected as unrealistic and it is realised that the collective behaviour of a group results from nonlinear interactions between its members. This is beginning to be understood. Dicks [22] quotes Joel Peck, an evolutionary theorist at the University of Sussex: 'There is no doubt that we were too hasty in trashing group selection. The theoretical models of the 60s and 70s were very oversimplified.' The same article discusses some fascinating experimental work of Swenson and Wilson [78] in which the unit of selection is expanded to an entire ecosystem. They discovered that it was possible to use selection on pond ecosystems, on the basis of ability to lower the acidity of water; or to improve the ability of soil ecosystems to digest pollutants by selecting at the ecosystem level.

Some caution is needed in interpreting their work, however. It is 'unnatural selection' since it is carried out by experimentalists using test tube ecosystems. The experimentalists get to choose what constitutes an ecosystem, and they *treat it as a group* for the process of selection – they either keep the test tube or they don't. (However, similar circumstances arguably do occur in nature.) A common mistake in this area is to use the term 'selection' as if it is independent of context. To see the error, imagine two species composed of individuals who are competing for some common resource. If the individuals in species P generally have an edge over those in species Q, then we expect to see species P surviving while Q is wiped out. In a sense, then, the two *species* have 'competed' and P has been 'selected'. However, neither word has the same meaning that it has when applied to individuals. Individuals in P are competing not just with individuals in Q, but with the other individuals in P; selection on the level of individuals can choose some members from P and some from Q – it is not an 'all or nothing' choice. Neither of these possibilities occur with 'competition' and 'selection' using P and Q as the only possible units.

How do we resolve these issues? Consider any system of organisms, be it a species or an ecosystem, as a complex system whose entities are organisms. The behaviour of the system is a consequence of interactions ('competition') between these organisms, resulting in some surviving to breed and others failing to do so ('selection'). This description is pure Darwin, but using a few modern buzzwords. Such a system can be viewed on many levels of description, however: we can focus on family lineages, species, and other kinds of group. If these groups are meaningful descriptors, we will observe various *emergent* patterns of behaviour: for example, over time one group may dominate and another may fall by the wayside. Such behaviour looks very much like 'selection' on the emergent level. However, the underlying complex system is *not* pitting one such group against another, with the rule that just one wins. It is pitting one organism against another: the group behaviour is an emergent consequence and on the group level many other things might happen instead of win/lose. So it is unwise to use the same word 'selection'. If there is a flaw in the interpretation of the experiments of Swenson and Wilson, it is that the 'selection' in these experiments is not an emergent consequence of selection for individuals: it is a direct choice on the level of ecosystems (test-tubes). This makes its implications for the earlier controversies debatable. Nevertheless, it represents an experimental breath of fresh air that blows away decades of sterile talk based on inadequate theoretical models. What we now need is a sensible conceptual framework for analysing group effects in evolution, and one necessary step is to recognise unexamined contextual differences in the meaning of terms like 'selection'.

The scenario described above can be related to these questions. We have argued that speciation is an emergent property of a complex system whose entities are organisms. These emergent patterns of phenotypic/genetic divergence can be observed on many levels of description, and this is the source of concepts such as kin selection, species selection, or ecosystem selection. We have just suggested that using a single term 'selection' on different levels is a mistake; now we can reinforce that message. For example, on the level of species, the win/lose dichotomy is not the only possible behaviour: one interesting new alternative is speciation.

We also see that the linear-theory objections to group selection, based on cheating by mavericks, do not apply in our scenario. To recap: we find that selection on individuals plus recombination leads to some finches with a predominance of type A alleles, and others with a predominance of Bs. Mixtures of A and B are eliminated. The usual arguments about mavericks have no force in these circumstances, for two reasons. One is that the interactions are nonlinear. The other is that the groups are defined by the phenotypes of the individuals, rather than the groups being set up in advance with all individuals being assumed to follow the common group behaviour. Each individual could in principle belong to either group, so moving from one group to the other is not 'cheating'. Mavericks, who subvert the process (within the confines of our model) by employing neither of the group strategies, simply get wiped out. To put it another way, the speciated state is dynamically stable.

Patchy Environments

We add one final remark about patchy environments. In the scenario we have described, potential genotypes of 'new' species are already present cryptically in the gene-pool before the speciation event itself happens. The divergence is triggered by environmental change. This has a consequence for patchy environments. Suppose the environment consists of a number of patches, only loosely coupled by interbreeding. Now imagine that some change in the environment occurs, perhaps with a general overall trend, but setting in at different times in different patches. As each patch experiences a change in environment that triggers speciation, there will be a rapid divergence of species in that patch. If two distinct patches undergo essentially the same environmental changes, the selective effects of environment will be much the same in both patches, and the pool of potential genomes will also be much the same. Therefore both patches will experience speciation, with much the same result – even on the *genetic* level. That is, we would expect to observe multiple, independent repetitions of the same speciation event in many locations. This is highly reminiscent of what has occurred in the evolution of African lake cichlids: see Meyer [59] and Stiassny and Meyer [75].

12. Conclusions

We have proposed a class of speciation models based on symmetric systems of ODEs posed in phenotypic space. These equations are nonlinear, so they can undergo symmetry-breaking bifurcations. Such bifurcations correspond to speciation events in the model. Models of this kind display certain typical kinds of bifurcation behaviour, independently of many details of the equations: we have discussed these 'model-independent' phenomena and used them to make predictions about sympatric speciation.

We have also described possible biological interpretations of the speciation process observed in the model, in terms of both phenotype and genotype, linking these to hypothetical models of beak sizes in birds and to field observations of various species, including grasshoppers and Darwin's finches.

Our main conclusions include:

- Speciation occurs in populations of *organisms*, not in a mean-field gene-pool.
- Genes render phenotypes fluid, and *recombination* is the dominant source of fluidity, not mutation (except in some asexual species, where recombination seldom if ever occurs – for example, in amazonogenesis).
- Speciation does not *require* environmental inhomogeneities, but can be assisted if they are present.
- Speciation does not *require* assortative mating, but can be stabilised rapidly by it.
- Divergence of phenotypes during speciation is the result of *selection*, not mutation.
- Mutations do play a role: they cryptically enlarge the gene-pool (genotypic space).
- It is the *distribution* of genomes in the genotypic space of *breeding adults* that matters, not statistical projections of that distribution, one allele at a time.
- The type of speciation that we have modelled is invisible to mean-field genetics. However, it can be detected by monitoring variance about the mean. Details, such as the number of divergent species, require more subtle observables. This is an area of considerable interest in field classification studies at the present time, and techniques based on symmetry detectives (Barany *et al.* [6]) might be applicable.
- Nonlinear effects create stable splits, and cause jumps/hysteresis, making the speciation process robust and 'irreversible'.
- Divergent phenotypes need not follow obvious changes in the distribution of 'niches' (but often do).
- Interchangeability of organisms within a species imply symmetry constraints on the dynamics, with significant effects on the typical behaviour.
- 'Founder populations' in the sympatric case are typically, though not universally, large (at least one third of the total population).
- Natural selection acts on organisms, not on genes. An evolutionary system is a complex system whose entities are the organisms. Speciation is an *emergent* consequence of selection on entities.

Appendix 1

We determine the S_N-invariant smooth functions $f : \mathbf{R}^N \to \mathbf{R}$ and S_N-equivariant smooth mappings $F : \mathbf{R}^N \to \mathbf{R}^N$. It is well known (see Golubitsky *et al.* [36]) that there is a Hilbert basis $\{\rho_1, \ldots, \rho_r\}$ of polynomial invariants, r finite, such that every smooth invariant can be expressed as $g(\rho_1, \ldots, \rho_r)$ for a smooth function $g : \mathbf{R}^n \to \mathbf{R}$ (Schwarz's Theorem). Furthermore, there is a finite set of polynomial equivariants $F_0, \ldots, F_s$ such that any smooth equivariant F can be expressed as

$$F = g_0 F_0 + \cdots + g_s F_s$$

for smooth invariants g_j (Poénaru's Theorem). Thus the smooth case reduces to the polynomial case.

Polynomial invariants for $\mathbf{S}_N$ on $\mathbf{R}^N$ have a Hilbert basis given by the *elementary symmetric functions*

$$
\begin{aligned}
\sigma_1 &= x_1 + \cdots + x_N \\
\sigma_2 &= x_1 x_2 + x_1 x_3 + \cdots + x_{N-1} x_N = \sum_{i \neq j} x_i x_j \\
&\cdots \\
\sigma_k &= \sum_{i_1, \ldots, i_k \text{ distinct}} x_{i_1} \ldots x_{i_k} \\
&\cdots \\
\sigma_N &= x_1 \ldots x_N
\end{aligned}
$$

or equivalently by sums of kth powers

$$
\pi_k = \sum_{i=1}^{N} x_i^k
$$

where $k = 1, \ldots, N$. For a proof see any text on Galois Theory, for example Stewart [74] Exercises 2.13 and 2.14. Note that the invariant polynomial $p(x) = 1$ is included.

The σ_k can also be defined by the identity

$$
(t - x_1)(t - x_2) \cdots (t - x_N) = t^N - \sigma_1 t^{N-1} + \sigma_2 t^{N-2} - \cdots \pm \sigma_N \qquad (16)
$$

for any indeterminate t. We shall require this identity in the proof of the next proposition.

Consider the $\mathbf{S}_N$-equivariants. The group $\mathbf{S}_N$ is generated by the N-cycle $\alpha = (1\,2\,3\ldots N)$ and the subgroup $\mathbf{S}_{N-1}$ consisting of all permutations of $\{2, \ldots, N\}$. Therefore a mapping is equivariant under $\mathbf{S}_N$ if and only if its is equivariant under $\mathbf{S}_{N-1}$ and α^{-1}. Suppose that $F : \mathbf{R}^N \to \mathbf{R}^N$ is an $\mathbf{S}_N$-equivariant polynomial mapping with components $F = (F_1, \ldots, F_N)$, and write F_1 in the form

$$
F_1(x) = \sum_d G_d(x_2, \ldots, x_N) x_1^d
$$

Equivariance under $\mathbf{S}_{N-1}$ implies that

$$
F_1(x) = \sum_d G_d(x_{\rho(2)}, \ldots, x_{\rho(N)}) x_1^d
$$

for all $\rho \in \mathbf{S}_{N-1}$, since $\mathbf{S}_{N-1}$ fixes the symbol 1. Therefore each G_d is *invariant* under $\mathbf{S}_{N-1}$. In addition, α^{-1}-equivariance implies that

$$
\begin{aligned}
F_i(x) &= \sum_d G_d(x_{i+1}, \ldots, x_{i-1}) x_i^d \\
&= \sum_d G_d(x_1, \ldots, x_{i-1}, x_{i+1}, \ldots, x_N) x_i^d
\end{aligned}
$$

For any polynomial $P(x_1, \ldots, x_N)$ we define

$$\lceil P(x_1, \ldots, x_N) \rceil = \begin{bmatrix} P(x_1, x_2, \ldots, x_{N-1}, x_N) \\ P(x_2, x_3, \ldots, x_N, x_1) \\ \vdots \\ P(x_N, x_1, \ldots, x_{N-2}, x_{N-1}) \end{bmatrix}$$

If P is $\mathbf{S}_{N-1}$-invariant, then $\lceil P \rceil$ is $\mathbf{S}_N$-equivariant, and conversely.

We now define $\mathbf{S}_N$-equivariants E_k, for $k = 0, 1, 2, \ldots$ by

$$E_k = \lceil x_1^k \rceil \tag{17}$$

Proposition 5. *The $\mathbf{S}_N$-equivariant polynomial mappings are generated over the $\mathbf{S}_N$-invariant polynomial functions by $E_0, \ldots, E_{N-1}$.*

Proof. Let p be any $\mathbf{S}_{N-1}$-invariant polynomial in $(x_2, \ldots, x_N)$, and denote its degree by k. We claim that there exists an $\mathbf{S}_N$-invariant polynomial $\tilde{p}$ in $(x_1, \ldots, x_N)$, whose degree is the same as that of p, and $\mathbf{S}_N$-invariant polynomials $q_1, \ldots, q_k$ in $(x_1, \ldots, x_N)$, such that

$$p = \tilde{p} + x_1 q_1 + \cdots + x_1^k q_k \tag{18}$$

To prove the claim, write p in the form

$$p(x) = G(\hat{\pi}_1, \ldots, \hat{\pi}_{N-1})$$

where $\hat{\pi}_l = x_2^l + \cdots + x_N^l$. (This is possible since the $\hat{\pi}_l$ for $l = 1, \ldots, N-1$ generate the $\mathbf{S}_{N-1}$-invariants.) Define

$$\tilde{p} = G(\pi_1, \ldots, \pi_{N-1})$$

Observe that by construction

$$\tilde{p}(0, x_2, \ldots, x_N) = p(x_2, \ldots, x_N)$$

Therefore, considered as a polynomial in x_1 with coefficients that are polynomials in $x_2, \ldots, x_N$, the difference $p - \tilde{p}$ is divisible by x_1 (using the remainder theorem, see for example Fraleigh [31] page 276). Expanding in powers of x_1, the claim follows, except that at the moment the q_j are only known to be $\mathbf{S}_{N-1}$-invariant. However, since p and $\tilde{p}$ have the same degree, all the q_j have smaller degree than p. Inductively, it follows that we can write p in the form (18) where now the q_j are $\mathbf{S}_N$-invariant polynomials in $(x_1, \ldots, x_N)$.

Suppose that $E(x)$ is equivariant, and write it in the form $E(x) = \lceil Q(x) \rceil$. Expanding Q as a polynomial in x_1 over the remaining variables, we can write

$$Q(x) = Q_0 + x_1 Q_1 + \cdots + x_1^k Q_k$$

for some k, where the Q_j are $\mathbf{S}_{N-1}$-invariants. By the claim above, we can rewrite Q in the form

$$Q(x) = R_0 + x_1 R_1 + \cdots + x_1^l R_l$$

where the R_j are $\mathbf{S}_N$-invariants. Therefore

$$
\begin{aligned}
E(x) &= \lceil Q(x) \rceil \\
&= \lceil R_0 + x_1 R_1 + \cdots + x_1^l R_l \rceil \\
&= R_0 \lceil 1 \rceil + R_1 \lceil x_1 \rceil + \cdots + R_l \lceil x_1^l \rceil \\
&= R_0 E_0 + R_1 E_1 + \cdots + R_l E_l
\end{aligned}
$$

so that the $\mathbf{S}_N$-equivariants are generated by the E_j over the $\mathbf{S}_N$-invariants.

Finally, we must show that only $E_0, \ldots, E_{N-1}$ are necessary as generators. This is a consequence of identity (16). Setting $t = x_1$ implies that

$$
x_1^N = \sigma_1 x_1^{N-1} - \sigma_2 x_1^{N-2} + \cdots \mp \sigma_N
$$

Repeatedly multiplying by x_1 and applying induction, it follows that for all $n \geq N$

$$
x_1^n = S_1 x_1^{N-1} + S_2 x_1^{N-2} + \cdots + S_N
$$

where the S_j are $\mathbf{S}_N$-invariants. Applying the operator $\lceil\ \rceil$ we see that E_n is a linear combination of $E_0, \ldots, E_{N-1}$ over the $\mathbf{S}_N$-invariants. $\qquad\square$

Appendix 2: Centre Manifold Reduction onto V_1

Equation (9) is an $\mathbf{S}_N$-equivariant ODE with 13 parameters. Here we describe work of Elmhirst [27] on the dynamics of the centre manifold reduction of such a cubic truncation to the space V_1, see also Cohen and Stewart [17]. This reduction contains all of the relevant bifurcation behaviour, but involves only 4 parameters, reducible to 3 by scaling. As such, it is much more tractable. Carr [14] contains general information on the process of centre manifold reduction.

We can obtain the general form of the centre manifold reduction onto V_1 by restriction and projection from the general $\mathbf{S}_N$-equivariant mapping on $\mathbf{R}^N$. To cubic order, we do this by imposing the relation $\pi_1 = 0$ on (8) and then projecting the results onto V_1. It is easy to check that only the following terms survive:

$$
G = b_2 E_1 + c_4 \left(E_2 - \frac{1}{N} \pi_2 E_0 \right) + d_5 \left(\pi_2 E_1 - \frac{1}{N} \pi_1 \pi_2 E_0 \right) + d_6 \left(E_3 - \frac{1}{N} \pi_3 E_0 \right)
$$

Cohen and Stewart [17] and Elmhirst [27] used different methods to obtain the equivalent form

$$
\begin{aligned}
G_i &= \lambda x_i + (N x_i^2 - \Sigma_2) + C(N x_i^3 - \Sigma_3) \\
&\quad + D(N x_i (x_1^2 + \cdots + x_{i-1}^2 + x_{i+1}^2 + \cdots + x_N^2) - \Sigma_{12})
\end{aligned} \tag{19}
$$

(after scaling x_i, λ by positive factors, which preserves orientation of the bifurcation diagram and stability and sets $c_4 = 1$), where

$$
\begin{aligned}
\Sigma_2 &= x_1^2 + \cdots + x_N^2 \\
\Sigma_3 &= x_1^3 + \cdots + x_N^3 \\
\Sigma_{12} &= \sum_{i \neq j} x_i^2 x_j
\end{aligned}
$$

For consistency with the calculations of Elmhirst [27] we work with (8) in §5. We also restrict attention to the case $c_4 = +1$. The case $c_4 = -1$ can be recovered by transforming x into $-x$, which turns the bifurcation diagrams upside-down, and we consider the case $c_4 = 0$ to be non-generic.

Appendix 3: Imperfect Symmetry

A common criticism of symmetry methods, especially in biology, is that real systems are seldom perfectly symmetric. Part of the answer is that the behavior of a dynamical system that is 'nearly symmetric' is much more like that of an idealised symmetric system than it is like the typical behavior of a completely asymmetric one. There are theoretical justifications of this assertion, such as the fact that 'normally hyperbolic' invariant sets of a dynamical system persist under small perturbations, Arrowsmith and Place [1]. Here we show, by numerical simulation, that two different ways to make the symmetry of the speciation model imperfect lead to much the same conclusions as the ideal symmetric model – even when both sources of imperfection are present together. These two ways are:

- Change the equations so that they are no longer perfectly symmetric in the N variables.
- Add stochastic terms to introduce a random element.

We show, by numerical experiment and theoretical discussion, that the main phenomena associated with (8) survive such modifications, so the quest for greater biological realism does not alter the main conclusions derived from the less realistic, but far more tractable, idealised equations.

We can break the $\mathbf{S}_N$ symmetry of (8) by making the coefficients vary slightly with the index i, and by replacing the terms $(x_1 + \cdots + x_N)$ and $(x_1^2 + \cdots + x_N^2)$ by $(r_1 x_1 + \cdots + r_N x_N)$ and $(s_1 x_1^2 + \cdots + s_N x_N^2)$ where the r_j and s_j differ slightly from 1. Here 'slightly' is governed by a new parameter g, which is typically 0.1 or thereabouts, indicating a 10% variation of the parameter values. In simulations, these variations are defined at the start of each run using a random number generator. The main change is that after bifurcation the traces do not converge as strongly, but they still clump. Figure 18 shows how Figure 9 changes under such perturbations: the change is minimal, even when the asymmetry is quite substantial.

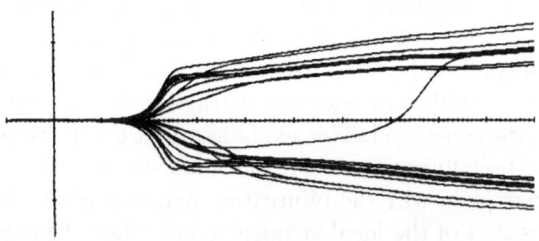

FIGURE 18. Imposed broken symmetry in (6).

An alternative way to introduce variability within species is to convert (8) into a stochastic model by adding Gaussian random noise (Brownian motion). In simulations we discretise and iterate:

$$x_i(t + \varepsilon) = x_i(t) + \varepsilon F(x(t)) + \sigma r_i(t)$$

Here ε is small (usually $\varepsilon = 0.0001$), σ determines the size of the noise, and r_i is a random variable distributed uniformly between -0.5 and 0.5. Figure 10 of §6 shows a typical bifurcation diagram for this case: as expected, it resembles noisy versions of Figure 9.

In fact there are general theoretical reasons for expecting the effects of these changes to the model to be small. For the addition of random terms, we can invoke the theory of stochastic nonlinear differential equations

$$dx = F(x, a)dt + \sigma dw \tag{20}$$

whose statistical behaviour is governed by the *Fokker-Planck equation*

$$\frac{\partial u}{\partial t} = \sigma \nabla^2 u - \nabla \cdot (uF)$$

Here the function $u = u(t)$ represents a time-varying smooth probability density, the term $\sigma \nabla^2 u$ represents random diffusion, and $\nabla \cdot (uF)$ represents the deterministic flow ($\sigma = 0$). Intuitively, solutions of (20) are like solutions of the deterministic equation subjected to random kicks at all instants of time, so for low noise ($\sigma \ll 1$) they look like slightly irregular versions of deterministic solutions. Formalising these ideas requires substantial technical effort: see Itô [47], McKean [60], Gihman and Skorohod [39]. It is known that as t tends to infinity, any solution of the Fokker-Planck equation converges to a unique steady state U. This implies that if F is Γ-equivariant for some group Γ and the noise is Γ-invariant then U must be Γ-invariant. In particular if for specific parameter values a the attractors of the deterministic system are hyperbolic equilibria and $\sigma \ll 1$, then U has peaks near all stable equilibria and is small everywhere else, see Kifer [51], Zeeman [83]. This result reflects the very long term behavior: because Brownian motion is long-range, the stochastic system can climb out of any potential well given enough time. However, that time scale is very long indeed, and what we see in simulations is more limited. Numerical evidence supports the conjecture that all attractors of (8) that bifurcate from the origin are equilibria, and these are generically hyperbolic: we assume this conjecture is true for the following discussion. In simulations, where the noise is taken to be short range (arguably more realistic for evolutionary biology), almost all trajectories eventually become trapped inside the basin of attraction of some equilibrium once the attraction towards that equilibrium exceeds the discretised noise level. This explains why the bifurcation behavior of the stochastic system resembles that of the deterministic one when $\sigma \ll 1$.

We can also explain why the bifurcation behavior of the deterministic modification resembles that of the ideal symmetric one when the amount g of imposed symmetry-breaking is small. The theoretical reason is normal hyperbolicity, which is a generic property in this context, Arrowsmith and Place [1].

Acknowledgements

We thank Marty Golubitsky for advice on equivariant bifurcation theory and ideas for further work, John Casti for suggesting links between our model and complex systems, and Johan Erlandson for information about hybrid zones in field observations.

References

[1] D.K. Arrowsmith and C.M. Place. *An Introduction to Dynamical Systems*, Cambridge U Press, Cambridge 1990.

[2] S.Y. Auyang. *Foundations of Complex-System Theories*, Cambridge U Press, Cambridge 1998.

[3] S.M. Baer, T. Erneux, and J. Rinzel. The slow passage through a Hopf bifurcation: delay, memory effects, and resonance, *SIAM J. Appl. Math.* **49** (1989) 55–71.

[4] C.R. Bantock and J.A Bayley. Visual selection for shell size in *Cepea* (Held), *J. Anim. Ecol.* **42** (1973) 247–261.

[5] C.R. Bantock, J.A. Bayley, and P.H. Harvey. Simultaneous selective predation on two features of a mixed sibling species population, *Evolution* **29** (1975) 636–649.

[6] E. Barany, M. Dellnitz, and M. Golubitsky. Detecting the symmetry of attractors, *Physica D* **67** (1993) 66–87.

[7] R.S.K. Barnes. Life-history strategies in contrasting populations of the coastal gastropod *Hydrobia* III. Lagoonal versus intertidal-marine *H. neglecta, Vie Milieu* **43** (1993) 73–83.

[8] R.S.K. Barnes. Investment in eggs in lagoonal *Hydrobia ventrosa* and life-history strategies in north-west European *Hydrobia* species, *J. Mar. Biol. Ass.* **74** (1994) 637–650.

[9] R.S.K. Barnes. Breeding, recruitment and survival in a mixed intertidal population of the mudsnails *Hydrobia ulvae* and *H. neglecta*, *J. Mar. Biol. Ass.* **76** (1996) 1003–1012.

[10] R.S.A. Beauchamp and P. Ullyett. Competitive relationships between certain species of fresh-water triclads, *J. Ecol.* **20** (1932) 200–208.

[11] R.K. Butlin and G.M. Hewitt. A hybrid zone between *Chorthippus parallelus parallelus* and *Chorthippus parallelus erythropus* (Orthoptera: Acrididae), *Biol. J. Linn. Soc.* **26** (1985) 287–299.

[12] B. Candelpergher, F. Diener, and M. Diener. Retard à la bifurcation: du local au global, in *Bifurcations of Planar Vector Fields, Luminy 1989*, Lecture Notes in Math. **1455**, Springer-Verlag, Berlin 1990, 1–19.

[13] J.Carr. *Applications of Centre Manifold Theory*, Springer-Verlag, New York 1981.

[14] C. Cicogna. Symmetry breakdown from bifurcation, *Lett. Nuovo Cimento* **31** (1981) 600–602.

[15] J. Cohen and I. Stewart. *The Collapse of Chaos*, Viking, New York 1994.

[16] J. Cohen and I. Stewart. *Figments of Reality*, Cambridge U Press, Cambridge 1997.

[17] J. Cohen and I. Stewart. Polymorphism viewed as phenotypic symmetry-breaking, in *Nonlinear Phenomena in Physical and Biological Sciences* (S.K. Malik ed.), Indian National Science Academy, New Delhi 2000, to appear.

[18] J. Cohen, I. Stewart, and T. Elmhirst. Symmetry, stochastics, and sympatric speciation, in preparation.

[19] L.M. Cook. *Coefficients of Natural Selection*, Hutchinson, London 1971.

[20] R. Dawkins. *The Selfish Gene*, Oxford U Press, Oxford 1976.

[21] R. Dawkins. *The Extended Phenotype*, Freeman, San Francisco 1982.

[22] L. Dicks. All for one! *New Scientist* (8 July 2000) 30–35.

[23] U. Dieckmann and M. Doebeli. On the origin of species by sympatric speciation, *Nature* **400** (1999) 354–457.

[24] F. Diener and M. Diener. Maximal delay, in *Dynamic Bifurcations, Luminy 1990*, Lecture Notes in Math. **1493** (ed. E. Benoit), Springer-Verlag, Berlin 1991, 71–86.

[25] Th. Dobzhansky. *Genetics and the Origin of Species*, Columbia U Press, New York 1937.

[26] N. Eldredge and S.J. Gould. Punctuated equilibrium: an alternative to phyletic gradualism, in *Models in Palaeobiology* (ed. T.J.M. Schopf), Cooper, San Francisco 1972.

[27] T. Elmhirst. *Symmetry-Breaking Bifurcations of S_N-Equivariant Vector Fields and Polymorphism*, MSc thesis, Mathematics Institute, U of Warwick 1998.

[28] T. Elmhirst. *Symmetry and Emergence in Polymorphism and Sympatric Speciation*, PhD Thesis, Mathematics Institute, University of Warwick, to appear.

[29] T. Fenchel. Factors determining the distribution patterns of mud snails (Hydrobiidae), *Oecologia* **20** (1975) 1–17.

[30] T. Fenchel. Character displacement and coexistence in mud snails, *Oecologia* **20** (1975) 19–32.

[31] J.B. Fraleigh. *A First Course in Abstract Algebra* (3rd. ed.), Addison-Wesley, Reading MA 1982.

[32] D. Francis, J. Diorio, D. Liu, and M.J. Meaney. Nongenomic transmission across generations of maternal behavior and stress responses in the rat, *Science* **286** (1999) 1155–1158.

[33] F. Galton. *Hereditary Genius*, Macmillan, London 1869.

[34] M. Golubitsky and D.G. Schaeffer. Bifurcations with **O**(3) symmetry including applications to the Bénard problem, *Commun. Pure Appl. Math.* **35** (1982) 81–111.

[35] M. Golubitsky and D.G. Schaeffer. *Singularities and Groups in Bifurcation Theory* vol 1, Applied Mathematical Sciences **51**, Springer-Verlag, New York 1985.

[36] M. Golubitsky, I.N. Stewart, and D.G. Schaeffer. *Singularities and Groups in Bifurcation Theory* vol 2, Applied Mathematical Sciences **69**, Springer-Verlag, New York 1988.

[37] P.R. Grant. Natural selection and Darwin's finches, *Scientific American* (October 1991) 60–65.

[38] P.R. Grant, B.R. Grant, J.N.M. Smith, I.J. Abbott, and L.K. Abbott. Darwin's finches: Population variation and natural selection, *Proc. Nat. Acad. Sci. USA* **73** (1976) 257–261.

[39] I.I. Gihman and A.V. Skorohod. *Stochastic Differential Equations*, Springer-Verlag, Berlin 1970.

[40] M.G. Hayes. *Geometric Analysis of Delayed Bifurcations*, PhD Thesis, Boston U 1999.

[41] M.K. Hecht and A. Hoffman. Why not neo-Darwinism? A critique of paleobiological challenges, in *Oxford Surveys in Evolutionary Biology* **3** (eds. R. Dawkins and M. Ridley), Oxford U Press, Oxford 1986, 1–47.

[42] M. Higashi, G. Takimoto, and N. Yamamura. Sympatric speciation by sexual selection, *Nature* **402** (1999) 523–526.

[43] J. Hofbauer and K. Sigmund. *The Theory of Evolution and Dynamical Systems*, Cambridge U Press, Cambridge 1988.

[44] R.B. Huey, G.W. Gilchrist, M.L. Carlson, D. Berrigan, and L. Serra. Rapid evolution of a geographic cline in size in an introduced fly, *Science* **287** (2000) 308–310.

[45] R.B. Huey and E.R. Pianka. Ecological character displacement in a lizard, *Am. Zool.* **14** (1974) 1127–1136.

[46] R.B. Huey, E.R. Pianka, M.E. Egan, and L.W. Coons. Ecological shifts in sympatry: Kalahari fossorial lizards (*Typhlosaurus*), *Ecology* **55** (1974) 304–316.

[47] K. Itô. On stochastic differential equations, *Mem. Amer. Math. Soc.* **4** (1961).

[48] S.A. Kauffman. *The Origins of Order*, Oxford U Press, Oxford 1993.

[49] S.A. Kauffman. *At Home in the Universe*, Viking, New York 1995.

[50] T.J. Kawecki. Sympatric speciation via habitat specialization driven by deleterious mutations, *Evolution* **51** (1997) 1751–1763.

[51] Y. Kifer. General random perturbations of hyperbolic and expanding transformations, *J. d'Anal. Math.* **47** (1986) 111–150.

[52] M. Kimura. *The Neutral Theory of Molecular Evolution*, Cambridge U Press, Cambridge 1983.

[53] A.S. Kondrashov and F.A. Kondrashov. Interactions among quantitative traits in the course of sympatric speciation, *Nature* **400** (1999) 351–354.

[54] D. Lack. *Darwin's Finches. An Essay on the General Biological Theory of Evolution*, Peter Smith, Gloucester MA 1968.

[55] R.C. Lewontin and J.L. Hubby. A molecular approach to the study of genic heterozygosity in natural populations II. Amount of variation and degree of heterozygosity in natural populations of *Drosophila pseudoobscura*, *Genetics* **54** (1966) 595–605.

[56] K. Mainzer. *Thinking in Complexity*, Springer-Verlag, Berlin 1994.

[57] E. Mayr. *Animal Species and Evolution*, Belknap Press, Cambridge MA 1963.

[58] E. Mayr. *Populations, Species, and Evolution*, Harvard U Press, Cambridge MA 1970.

[59] A.Meyer. Phylogenetic relationships and evolutionary processes in East African cichlid fishes, *Trends in Ecol. and Evol.* **8** (1993) 279–284.

[60] H.P. McKean. *Stochastic Integrals*, Academic Press, New York 1969.

[61] R.M. Neems and R.K. Butlin. Divergence in cuticular hydrocarbons between parapatric subspecies of the meadow grasshopper *Chorthippus parallelus* (Orthoptera, Acrididae), *Biol. J. Linn. Soc.* **54** (1995) 139–149.

[62] A.I. Niehstadt. Persistence of stability loss for dynamical bifurcations I, *Diff. Eq.* **23** (1987) 1385–1391.

[63] A.I. Niehstadt. Persistence of stability loss for dynamical bifurcations II, *Diff. Eq.* **24** (1988) 171–196.

[64] E. Pennisi. Nature steers a predictable course, *Science* **287** (2000) 207–208.

[65] R. Rice and E.E. Hostert. Laboratory experiments on speciation: what have we learned in 40 years? *Evolution* **47** (1993) 1637–1653.

[66] M. Ridley. *The Problems of Evolution.* Oxford U Press, Oxford 1985.

[67] M. Ridley. *Evolution,* Blackwell, Oxford 1996.

[68] C.D. Rollo. *Phenotypes,* Chapman and Hall, London 1995.

[69] H.D. Rundle, L. Nagel, J.W. Boughman, and D. Schluter. Natural selection and parallel speciation in sympatric sticklebacks, *Science* **287** (2000) 306–308.

[70] S.N. Salthe. *Evolutionary Biology,* Holt, Rinehart and Winston, New York 1972.

[71] S.L. Reilly, R.E. Ferrell, B.A. Kottke, M.I. Kamboh, and C.F. Sing. The gender-specific apolipoprotein E genotype influence on the distribution of plasma lipids and apolipoproteins in the population of Rochester, Minnesota. I. Pleiotropic effects on means and variances. *Am. J. Hum. Genet.* **49** (1991) 1155–1166.

[72] S.L. Reilly, R.E. Ferrell, B.A. Kottke, and C.F. Sing. The gender-specific apolipoprotein E genotype influence on the distribution of plasma lipids and apolipoproteins in the population of Rochester, Minnesota. II. Regression relationships with concomitants. *Am. J. Hum. Genet.* **51** (1992) 1311–1324.

[73] S.L. Reilly, R.E. Ferrell, and C.F. Sing. The gender-specific apolipoprotein E genotype influence on the distribution of plasma lipids and apolipoproteins in the population of Rochester, Minnesota. III. Correlations and covariances. *Am. J. Hum. Genet.* **55** (1994) 1001–1018.

[74] I. Stewart *Galois Theory* (2nd ed.), Chapman & Hall, London 1989.

[75] M.L.J. Stiassny and A. Meyer. Cichlids of the Rift Lakes, *Scientific American* (February 1999) 64–69.

[76] J. Su. Delayed oscillation phenomena in the Fitzhugh-Nagumo equation, *J. Diff. Eq.* **105** (1993) 180–215.

[77] J. Su. Effects of periodic forcing on delayed bifurcations, *J. Dyn. Diff. Eq.* **9** (1997) 561–625.

[78] W. Swenson and D.S. Wilson, Artificial ecosystem selection, *Proc. Nat. Acad. Sci.,* to appear.

[79] T. Tregenza and R.K. Butlin. Speciation without isolation, *Nature* **400** (1999) 311–312.

[80] A. Vanderbauwhede, *Local Bifurcation and Symmetry,* Habilitation Thesis, Rijksuniversiteit Gent, 1980, (also Res. Notes Math. **75**, Pitman, Boston 1982).

[81] K. Winker. Migration and speciation, *Nature* **404** (2000) 36.

[82] A.P. Wolffe and M.A. Matzke. Epigenetics: regulation through repression, *Science* **286** (1999) 481–486.

[83] E.C. Zeeman. Stability of dynamical systems, *Nonlinearity* **1** (1988) 115–155.

Trends in Mathematics:
Bifurcations, Symmetry and Patterns, 55–66
© 2003 Birkhäuser Verlag Basel/Switzerland

Bifurcation and Planar Pattern Formation for a Liquid Crystal

Martin Golubitsky and David Chillingworth

Abstract. We consider the Landau – de Gennes model for the free energy of a liquid crystal, and discuss the geometry of its equilibrium set (critical points) for spatially uniform states in the absence of external fields. Using equivariant bifurcation theory we classify (on the basis of symmetry considerations independent of the model) square and hexagonally periodic patterns that can arise when a homeotropic nematic state becomes unstable, perhaps as a consequence of an applied magnetic or electric field.

1. Introduction

In the Landau theory of phase transitions for a liquid crystal the degree of coherence of alignment of molecules is usually represented by a field of symmetric 3×3 tensors $Q(\mathbf{x})$, $\mathbf{x} \in \mathbf{R}^3$ with trace $\mathrm{tr}(Q) = 0$ (the *tensor order parameter*) [15]. We think of Q as the second moment of a probability distribution for the directional alignment of a rod-like molecule. In a spatially uniform system, Q is independent of $\mathbf{x} \in \mathbf{R}^3$. When $Q = 0$ the system is *isotropic*, with molecules not aligned in any particular direction. If there is a preferred direction along which the molecules tend to lie (but with no positional constraints) the liquid crystal is in *nematic* phase. There are many other types of phase involving local and global structures, see [15].

SYMMETRIES IN THE ORDER PARAMETER. The complex linear space V of traceless symmetric 3×3 matrices Q is 5-dimensional over $\mathbf{C}$ with unitary basis

$$\frac{1}{2}\{M_0, M_{\pm 1}, M_{\pm 2}\}$$

where

$$M_0 = \sqrt{\tfrac{2}{3}}\begin{pmatrix} -1 & 0 & 0 \\ 0 & -1 & 0 \\ 0 & 0 & 2 \end{pmatrix} \quad M_{\pm 1} = \begin{pmatrix} 0 & 0 & \pm 1 \\ 0 & 0 & i \\ \pm 1 & i & 0 \end{pmatrix} \quad M_{\pm 2} = \begin{pmatrix} 1 & \pm i & 0 \\ \pm i & -1 & 0 \\ 0 & 0 & 0 \end{pmatrix}.$$

A state (phase) of the liquid crystal in $\mathbf{R}^3$ is given by the real part of a map $Q : \mathbf{R}^3 \to V$. At each point $\mathbf{x}$ in space the rod-like molecule is assumed to align along the eigendirection corresponding to the largest eigenvalue of $Q(\mathbf{x})$.

The action of rigid motions in $\mathbf{R}^3$ on a state is defined as follows. Let $\gamma \in \mathbf{O}(3)$ and let $T_{\mathbf{y}}$ be translation by $\mathbf{y} \in \mathbf{R}^3$. Then

$$\begin{aligned}
(T_{\mathbf{y}}Q)(\mathbf{x}) &= Q(\mathbf{x} - \mathbf{y}) \\
(\gamma \cdot Q)(\mathbf{x}) &= \gamma Q(\gamma^{-1}\mathbf{x})\gamma^{-1}.
\end{aligned} \tag{1.1}$$

That is, translations just translate $Q(\mathbf{x})$ while rotations and reflections act simultaneously by rigid motion on the domain of $Q(\mathbf{x})$ and by conjugacy in the range.

THE FREE ENERGY FORMULATION. Equilibrium states of the liquid crystal (ignoring boundary effects which in physical situations do play a crucial role) are taken to be critical points of a smooth real-valued *free energy* functional

$$F(Q) = \frac{1}{\text{Vol}} \int \mathcal{F}(Q(\mathbf{x}))d\mathbf{x}$$

defined for real Q, where the free energy density $\mathcal{F}$ is invariant under the Euclidean action (1.1). A standard free energy is given by the Landau-de Gennes model [9]

$$\begin{aligned}
\mathcal{F}(Q) &= \tfrac{1}{2}\tau \operatorname{tr}(Q^2) - \tfrac{1}{3}B\operatorname{tr}(Q^3) + \tfrac{1}{4}C\big(\operatorname{tr}(Q^2)\big)^2 \\
&\quad + C_1|\nabla Q|^2 + C_2|\nabla \cdot Q|^2 - 2DQ \cdot \nabla \wedge Q
\end{aligned} \tag{1.2}$$

where B, C, C_1, C_2, D are constants of the material and τ represents deviation from a critical temperature. The notation here is

$$|\nabla Q|^2 = \sum_{i,j}|\nabla Q_{ij}|^2,$$

$$|\nabla \cdot Q|^2 = \sum_{j}|\nabla \cdot Q_j|^2,$$

$$Q \cdot \nabla \wedge Q = \sum_{j}Q_j \cdot \nabla \wedge Q_j$$

where Q_j is the jth column of Q. This is a general $\mathbf{O}(3)$-invariant function of degree at most four in Q [8] and at most two in the first-order spatial derivatives of Q.

In this paper we discuss aspects of bifurcations of spatially homogeneous states (Section 2) and spatially periodic nematic liquid crystal states (Section 3).

2. Spatially uniform equilibrium states

For a spatially uniform state the derivatives of Q are zero and we are reduced to considering critical points of $F : V \to \mathbf{R}$ restricted to real matrices. Symmetry implies that every equilibrium state corresponds to a group orbit of equilibria. Since every symmetric matrix can be diagonalized by an orthogonal matrix, it follows that every group orbit of equilibria contains a diagonal trace zero matrix. Thus, to study bifurcation of equilibria, we can restrict attention to the 2-dimensional

τ range	equilibria	stability
$\tau > \tau_0$	$Q = 0$	stable
$\tau_0 > \tau > 0$	$Q = 0$	stable
	$Q = Q_i := \eta_i Q(0,1)$	$i = 1$ stable
	$\eta_1 < \eta_2 < 0$	
	$Q = \widetilde{Q}_i := -\frac{1}{2}\eta_i Q(\pm\sqrt{3},1)$	$i = 2$ unstable
$0 > \tau$	$Q = 0$	unstable
	$Q = Q_1, \widetilde{Q}_1, \quad \eta_1 < 0$	stable
	$Q = Q_2, \widetilde{Q}_2, \quad \eta_2 > 0$	unstable

TABLE 1. Equilibria as a function of τ where $\tau_0 = \frac{B^2}{24C}$ and $Q(\rho, \eta)$ is defined in (2.1).

space of *diagonal* traceless matrices as in [8, XV,§6]. We express such Q in the form

$$Q = Q(\rho, \eta) = -\eta\sqrt{\frac{3}{2}}M_0 + \rho\frac{\sqrt{3}}{2}(M_2 + M_{-2}) \tag{2.1}$$

since, in these coordinates,

$$\mathrm{tr}(Q^2) \;=\; 6(\rho^2 + \eta^2) \tag{2.2}$$
$$\mathrm{tr}(Q^3) \;-\; 6\eta(3\rho^2 - \eta^2) = 6\,\mathrm{Im}(\rho + \imath\eta)^3. \tag{2.3}$$

The function F restricted to the space $U \cong \mathbf{R}^2$ of matrices (2.1) is invariant with respect to the action of $\mathbf{D}_3$ in the (ρ, η)-plane generated by rotation by $2\pi/3$ and reflection in the η-axis. Thus

Proposition 2.1. *Every nonzero critical orbit of F meets U in a $\mathbf{D}_3$-symmetric configuration.* □

By first considering the restriction of F to the η-axis and then exploiting symmetry it is straightforward to deduce the description of equilibria for the system $\dot{Q} = -\mathrm{grad}\,F(Q)$ on U given in Table 1.

As τ decreases through $\tau_0 = \frac{B^2}{24C}$ there are simultaneous saddle-node creations of pairs of equilibria at $(\rho, \eta) = (0, \eta_0)$ and $-\frac{1}{2}\eta_0(\pm\sqrt{3}, 1)$ where $\eta_0 = -2\tau_0/B$; subsequently the innermost equilibria approach the origin and coalesce at a degenerate critical point there as τ decreases to 0, emerging on the other side as τ becomes negative. See Figures 1 and 2.

The physical interpretation is that for $\tau > \tau_0$ the only stable phase is isotropic ($Q = 0$) while for $\tau < \tau_0$ there are further stable nematic phases with molecules aligned in a particular direction: any one alignment has the same free energy as any other. The isotropic phase loses stability when τ becomes negative. This familiar transition is described for example in [15].

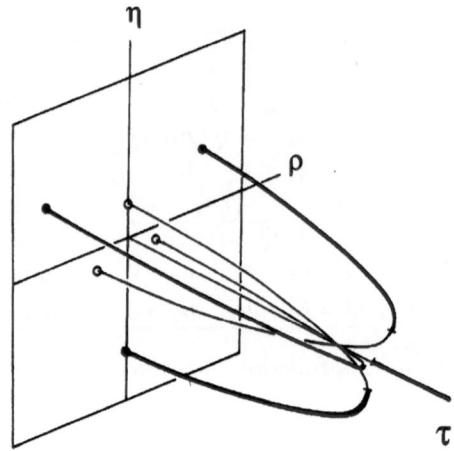

FIGURE 1. Bifurcation diagram for critical points of grad F on U.

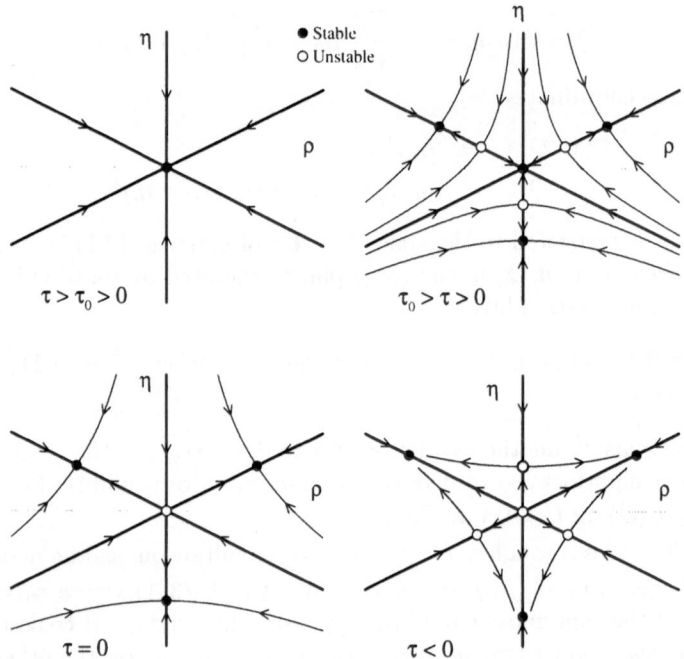

FIGURE 2. Schematic phase portraits for $\dot{Q} = -\operatorname{grad} F(Q)$.

REMARKS ON EQUILIBRIA IN A SHEAR FLOW. In the presence of a constant shear
flow the system loses its $\mathbf{O}(3)$ symmetry and most of the equilibria disappear.

However, some do remain, along with a variety of interesting dynamical phenomena including Hopf bifurcations, Takens-Bogdanov bifurcation and period-doubling that come into play as τ decreases [14, 16]. A rigorous geometric analysis of some of these phenomena is given in [3], where it is shown that all equilibria are invariant under reflection in the plane of the shear flow, with the curious exception of a continuum (ellipse) of out-of-plane equilibria that arise with codimension 1. This non-generic behavior casts doubt on the robustness (structural stability) of the Landau-de Gennes model in the presence of a shear flow.

3. Spatially periodic equilibrium states

Suppose a spatially uniform equilibrium Q_0 loses stability to a spatially periodic state. In this section we use group representation theory (following [8, 7, 4]) to extract information about nonlinear behavior at bifurcation that is independent of the model.

Specifically, we consider local bifurcation from a planar layer of a homeotropic nematic liquid crystal Q_0 that is assumed to have constant alignment in the vertical direction to one that has spatially varying alignment in the planar directions. We assume that the new states are spatially periodic with respect to some planar lattice. The symmetry group for this discussion is the planar Euclidean group rather than the Euclidean group in three dimensions, as in the previous sections.

The fact that liquid crystals can display spatial periodicity with respect to a planar lattice is well known by experiment. For example, Figure 3 illustrates a so-called prewavy pattern [13, 10] while Figure 4 shows two types of chevron [10]. (We are grateful to the authors of the abstracts [11],[12] for these pictures.) Several striking photographs of periodic patterns can also be found in [6].

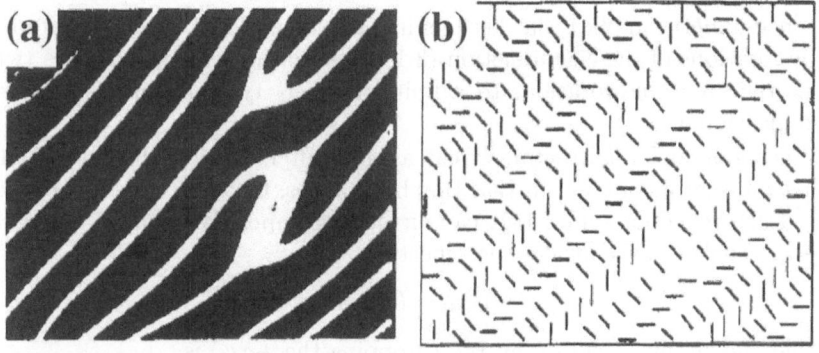

FIGURE 3. Prewavy pattern seen with crossed polarizers (a), and corresponding director field (b). The scale bar is 200μm. (Courtesy J.-H.Huh.)

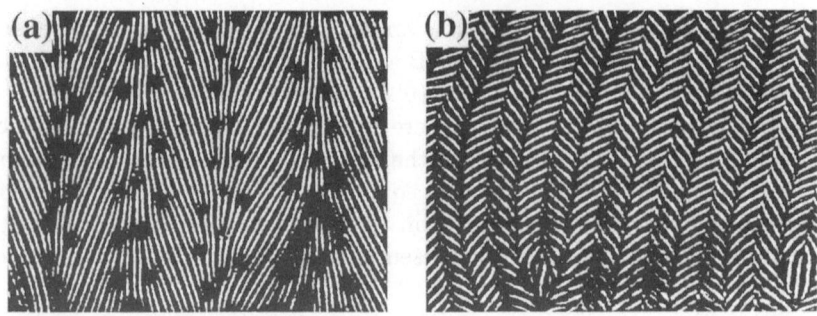

FIGURE 4. Two types of chevrons: (a) defect-mediated, (b) defect-free. The scale bar is $200\mu m$. (Courtesy J.-H.Huh.)

Linear Theory

Let L denote the linearization of the governing system of PDEs at Q_0 (for the free energy model we have $L = D^2 F(Q_0)$). Bifurcation occurs at parameter values where L has a nonzero kernel. Planar translation symmetry implies that eigenfunctions of L have the plane wave form

$$e^{2\pi i \mathbf{k} \cdot \mathbf{x}} Q + c.c. \tag{3.1}$$

where $Q \in V$ is a constant matrix and $\mathbf{k} \in \mathbf{R}^2$ is a wave vector. Let

$$W_{\mathbf{k}} = \{e^{2\pi i \mathbf{k} \cdot \mathbf{x}} Q + c.c. : Q \in V\} \tag{3.2}$$

be the ten-dimensional L-invariant real linear subspace consisting of such functions.

Rotations and reflections $\gamma \in \mathbf{O}(2)$ act on $W_{\mathbf{k}}$ by

$$\gamma(e^{2\pi i \mathbf{k} \cdot \mathbf{x}} Q) = e^{2\pi i (\gamma \mathbf{k}) \cdot \mathbf{x}} \gamma Q \gamma^{-1}. \tag{3.3}$$

Rotation symmetry implies that ker L is infinite-dimensional, since it contains all possible rotations of the eigenfunction (3.1). Restricting to planar lattices (which restricts the allowable rotations to a finite number) typically makes the kernel finite-dimensional.

When looking for nullvectors we can assume, after rotation, that $\mathbf{k} = k(1, 0, 0)$ and that nullvectors of L are in $W_{\mathbf{k}}$. Bosch Vivancos, Chossat, and Melbourne [1] observed that reflection symmetries can further decompose $W_{\mathbf{k}}$ into two L-invariant subspaces. To see why, consider the reflection

$$\kappa(x, y, z) = (x, -y, z).$$

Note that the action (3.3) of κ on $W_{\mathbf{k}}$ (dropping the $+c.c.$) is

$$\kappa\left(e^{2\pi i \mathbf{k} \cdot \mathbf{x}} Q\right) = e^{2\pi i \kappa(\mathbf{k}) \cdot \mathbf{x}} \kappa Q \kappa^{-1} = e^{2\pi i \mathbf{k} \cdot \mathbf{x}} \kappa Q \kappa^{-1}.$$

Since $\kappa^2 = 1$, the subspace $W_{\mathbf{k}}$ itself decomposes as

$$W_{\mathbf{k}} = W_{\mathbf{k}}^+ \oplus W_{\mathbf{k}}^- \tag{3.4}$$

where κ acts trivially on $W_{\mathbf{k}}^+$ and as minus the identity on $W_{\mathbf{k}}^-$. We call functions in $W_{\mathbf{k}}^+$ *even* and functions in $W_{\mathbf{k}}^-$ *odd*. Bifurcations based on even eigenfunctions are called *scalar* and bifurcations based on odd eigenfunctions are called *pseudoscalar*.

To determine the form of the scalar and pseudoscalar matrices (that is, those matrices Q^+, Q^- appearing in $W_{\mathbf{k}}^+$ and $W_{\mathbf{k}}^-$ respectively), we need to compute the effect of conjugacy by $\kappa \in \mathbf{O}(3)$ on V. The subspace of V where κ acts as the identity is

$$V^+ = \text{span}\{\, M_0,\ M_1 - M_{-1},\ M_2 + M_{-2}\,\}$$

and the space where κ acts as minus the identity is

$$V^- = \text{span}\{\, M_1 + M_{-1},\ M_2 - M_{-2}\,\}.$$

A further simplification can be made. Consider $R_\pi \in \mathbf{O}(3)$ given by $(x, y, z) \mapsto (-x, -y, z)$. Since (dropping the $+ c.c.$)

$$R_\pi(Q e^{2\pi i \mathbf{k} \cdot \mathbf{x}}) = R_\pi \cdot Q\, e^{2\pi i \mathbf{k} \cdot R_\pi(\mathbf{x})} = R_\pi \cdot Q\, e^{-2\pi i \mathbf{k} \cdot \mathbf{x}} = \overline{R_\pi \cdot Q}\, e^{2\pi i \mathbf{k} \cdot \mathbf{x}}$$

the associated action of R_π on V is related to the conjugacy action by

$$R_\pi(Q) = \overline{R_\pi \cdot Q}.$$

Since L commutes with R_π and $R_\pi^2 = 1$, the subspaces of the kernel of L where $R_\pi(Q) = Q$ and $R_\pi(Q) = -Q$ are L-invariant. Therefore, we can assume that Q is in one of these two subspaces. Note that translation by $\ell = \frac{1}{4}\mathbf{k}/k^2$ implies that if $v(\mathbf{x}, Q) = e^{2\pi i \mathbf{k} \cdot \mathbf{x}} Q$ is an eigenfunction then $iv(\mathbf{x}, Q)$ is a (symmetry related) eigenfunction. It follows from (3) that if R_π acts as minus the identity on Q, then R_π acts as the identity on iQ. Thus we can assume without loss of generality that

$$R_\pi(Q) = \overline{Q},$$

that is, Q is R_π-invariant. Therefore we have proved

Lemma 3.1. *Generically eigenfunctions in $V_{\mathbf{k}}$ have the form $e^{2\pi i \mathbf{k} \cdot x} Q + c.c.$ where Q is nonzero, R_π-invariant, and either even or odd.* $\qquad\square$

Lemma 3.1 implies that typically eigenspaces are two-dimensional subspaces of $W_{\mathbf{k}}^+$ or $W_{\mathbf{k}}^-$ and have the form

$$V_{\mathbf{k}}^+ = \{z e^{2\pi i \mathbf{k} \cdot \mathbf{x}} Q^+ : z \in \mathbf{C}\}$$
$$V_{\mathbf{k}}^- = \{z e^{2\pi i \mathbf{k} \cdot \mathbf{x}} Q^- : z \in \mathbf{C}\}$$

where Q^+ and Q^- are R_π-invariant. We check easily that

$$R_\pi \cdot M_0 = \overline{M}_0, \quad R_\pi \cdot M_{\pm 1} = M_{\mp 1}, \quad R_\pi \cdot M_{\pm 2} = M_{\mp 2}$$

and so by R_π-invariance we may assume that

$$\begin{aligned} Q^+ &= aM_0 + b(M_2 + M_{-2}) + ic(M_1 - M_{-1}), & a, b, c \in \mathbf{R} \\ Q^- &= g(M_1 + M_{-1}) + ih(M_2 - M_{-2}), & g, h \in \mathbf{R} \end{aligned} \qquad (3.5)$$

where $a, b, c, g, h \in \mathbf{R}$ are specific values chosen by L (cf. [7, §5.7]).

The Planforms

We now consider 2-dimensional patterns by disregarding the z-coordinate in $\mathbf{x}$ (but not in Q) and restrict attention to equilibrium states that are periodic with respect to a square or hexagonal lattice in the x,y-plane.

THE SQUARE LATTICE. The holohedry (the rotations and reflections that preserve the lattice) is $\mathbf{D}_4$ generated by κ and ξ, where ξ is counterclockwise rotation of the plane by $\frac{\pi}{2}$. We study the case where the critical dual wave vectors have shortest length and the kernel of L is four-dimensional:

$$V_{\mathbf{k}}^+ \oplus \xi\left(V_{\mathbf{k}}^+\right) \qquad \text{or} \qquad V_{\mathbf{k}}^- \oplus \xi\left(V_{\mathbf{k}}^-\right).$$

Therefore, we can write the general eigenfunction in the scalar case as

$$R^+(\mathbf{x}) = z_1 e^{2\pi i \mathbf{k}_1 \cdot \mathbf{x}} Q^+ + z_2 e^{2\pi i \mathbf{k}_2 \cdot \mathbf{x}} \xi Q^+ \xi^{-1} + c.c. \tag{3.6}$$

and in the pseudoscalar case as

$$R^-(\mathbf{x}) = z_1 e^{2\pi i \mathbf{k}_1 \cdot \mathbf{x}} Q^- + z_2 e^{2\pi i \mathbf{k}_2 \cdot \mathbf{x}} \xi Q^- \xi^{-1} + c.c. \tag{3.7}$$

In each case there are two axial subgroups (isotropy subgroups with 1-dimensional fixed-point spaces, that we call axial directions), so the equivariant branching lemma [8, 4, 7] predicts that bifurcations from a spatially uniform nematic state will occur along these axial directions at least. Up to conjugacy by an element of $\mathbf{D}_4 \dotplus \mathbf{T}^2$, the direction $(z_1, z_2) = (1, 0)$ corresponds to rolls and the direction $(z_1, z_2) = (1, 1)$ corresponds to squares.

To visualize the patterns of bifurcating solutions we assume a layer of liquid crystal material in the x, y-plane, possibly with an applied magnetic field in the z direction. We assume that the initial solution corresponds to a nematic phase with all molecules oriented in the z direction and that a symmetry-breaking bifurcation occurs as the strength of the magnetic field, temperature or other parameter is decreased. At each point (x, y) we choose the eigendirection corresponding to the largest eigenvalue of the symmetric 3×3 matrix $Q(\mathbf{x})$ at $\mathbf{x} = (x, y)$ and we plot only the x, y components of that line field. In this picture, a line element that degenerates to a point corresponds to a vertical eigendirection, so the initial solution looks like at array of points. In Figures 5 and 6 we plot solutions corresponding to scalar and pseudoscalar rolls and squares. Note that pseudoscalar rolls form a chevron pattern that can be compared to Figure 4(b).

THE HEXAGONAL LATTICE. The holohedry is $\mathbf{D}_6$ and is generated by κ and ξ, where ξ is counterclockwise rotation of the plane by $\frac{\pi}{3}$. The action of ξ on Q is

$$\xi(Q) = \xi Q \xi^{-1}.$$

On the hexagonal lattice, we also study the case where the dual wave vectors have shortest length and the kernel of L is six-dimensional. The dual wave vectors can be chosen to be

$$\mathbf{k}_1 = (1, 0) \qquad \mathbf{k}_2 = \tfrac{1}{2}(-1, \sqrt{3}) \qquad \mathbf{k}_3 = \tfrac{1}{2}(-1, -\sqrt{3}).$$

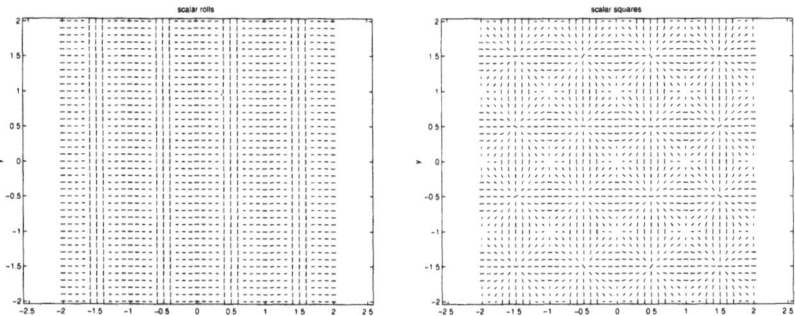

FIGURE 5. Square lattice with scalar representation: (left) rolls; (right) squares.

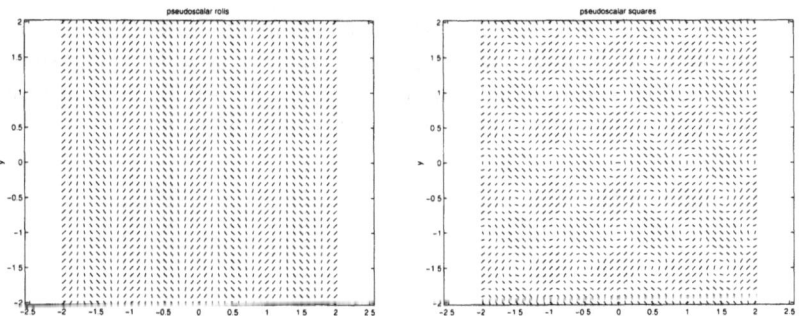

FIGURE 6. Square lattice with pseudoscalar representation: (left) anti-rolls; (right) anti-squares.

The eigenspaces are then

$$V_{\mathbf{k}}^{+} \oplus \xi^{2}\left(V_{\mathbf{k}}^{+}\right) \oplus \xi^{4}\left(V_{\mathbf{k}}^{+}\right) \quad \text{or} \quad V_{\mathbf{k}}^{-} \oplus \xi^{2}\left(V_{\mathbf{k}}^{-}\right) \oplus \xi^{4}\left(V_{\mathbf{k}}^{-}\right).$$

Therefore, we can write the general eigenfunction in the scalar case as

$$z_{1}e^{2\pi i \mathbf{k}_{1}\cdot\mathbf{x}}Q^{+} + z_{2}e^{2\pi i \mathbf{k}_{2}\cdot\mathbf{x}}\xi^{2}Q^{+}\xi^{4} + z_{3}e^{2\pi i \mathbf{k}_{3}\cdot\mathbf{x}}\xi^{4}Q^{+}\xi^{2} + c.c.$$

and in the pseudoscalar case as

$$z_{1}e^{2\pi i \mathbf{k}_{1}\cdot\mathbf{x}}Q^{-} + z_{2}e^{2\pi i \mathbf{k}_{2}\cdot\mathbf{x}}\xi^{2}Q^{-}\xi^{4} + z_{3}e^{2\pi i \mathbf{k}_{3}\cdot\mathbf{x}}\xi^{4}Q^{-}\xi^{2} + c.c.$$

It is well-known from analyses of Bénard convection (see [8]) that on the scalar hexagonal lattice there are two branches of axial solutions – hexagons and rolls – and that the hexagons come in two types hexagons^{+} and hexagons^{-}. For rolls we may take $(z_{1}, z_{2}, z_{3}) = (1, 0, 0)$ and for hexagons^{+} and hexagons^{-} we may take $(z_{1}, z_{2}, z_{3}) = \pm(1, 1, 1)$. Sample hexagon planforms are shown in Figure 7. Rolls are the same as those in Figure 5.

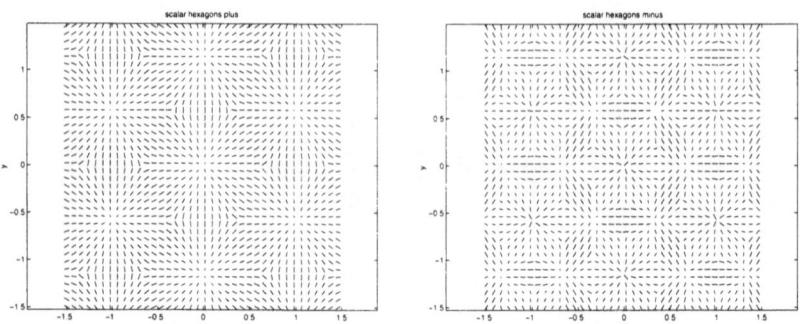

FIGURE 7. Hexagons on hexagonal lattice with scalar representation: (left) hexagons$^+$; (right) hexagons$^-$.

Bosch Vivancos *et al.* [1] and Bressloff *et al.* [2] show that in the pseudoscalar representation hexagons are given by $(z_1, z_2, z_3) = (1, 1, 1)$, triangles by $(z_1, z_2, z_3) = (i, i, i)$, and rectangles by $(z_1, z_2, z_3) = (1, -1, 0)$. Rolls are the same as in Figure 6. The remaining planforms are shown in Figure 8.

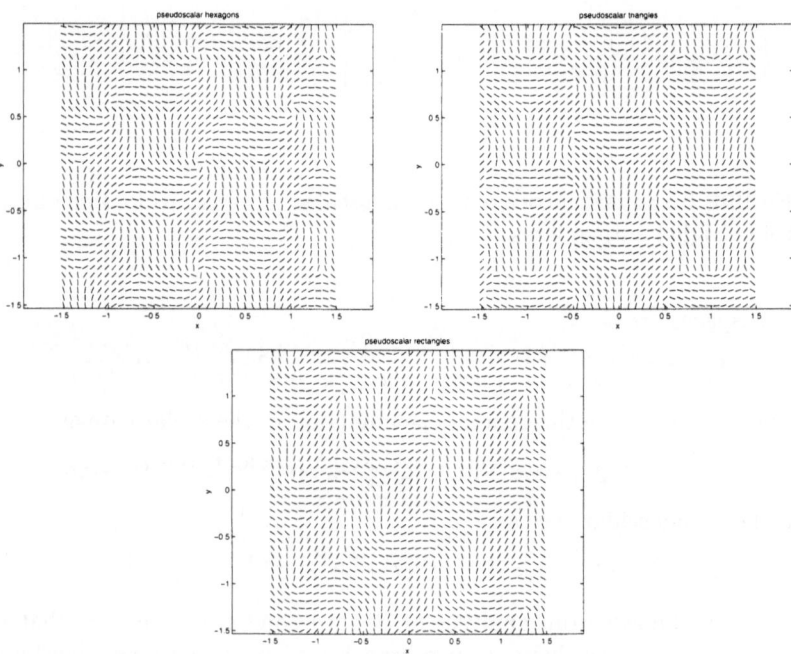

FIGURE 8. Hexagonal lattice with pseudoscalar representation: (upper left) hexagons; (upper right) triangles; (lower) rectangles.

Free energy interpretations

These results imply that there are two types of steady-state bifurcations, scalar and pseudoscalar, that can occur from a spatially homogeneous equilibrium to spatially periodic equilibria. If a scalar bifurcation occurs, then generically all of the scalar planforms that we listed (rolls, squares, hexagons$^+$, hexagons$^-$) will be solutions. Similarly, if a pseudoscalar bifurcation occurs, then generically all of the planforms that we listed (anti-rolls, anti-squares, hexagons, triangles, rectangles) will be solutions. We have not discussed the difficult issue of stability of these solutions.

What remains is to complete a linear calculation to determine when a steady-state bifurcation occurs and whether it is scalar or pseudoscalar. The outline of such a calculation goes as follows. We need to compute a dispersion curve for both scalar and pseudoscalar eigenfunctions. That is, for each wave length $k = |\mathbf{k}|$ we must determine the first value of the bifurcation parameter λ where L has a nonzero kernel. Call that value λ_k. The curve (k, λ_k) is called the *dispersion curve*. We then find the minimum value $\lambda_* = \lambda_{k_*}$ on the dispersion curve; the corresponding wave length k_* is the *critical* wave length. We expect the first instability of the spatially homogeneous equilibrium to occur at the value λ_* of the bifurcation parameter.

In principle, these calculations can be completed for the model equations (1.2) or a similar model, extending related calculations for bifurcation from the isotropic phase carried out by [9]. We defer the completion of this task to a future paper.

Acknowledgements

We are grateful to Tim Sluckin for instructive conversations on the mathematical physics of liquid crystals and to Ian Melbourne and Ian Stewart for discussions concerning pseudoscalar representations. We also thank Eva Vicente Alonso for Figure 2 and J.-H.Huh for permission to use the pictures in Figures 3 and 4. The research of MG was supported in part by NSF Grant DMS-0071735.

References

[1] I. Bosch Vivancos, P. Chossat, and I. Melbourne. New planforms in systems of partial differential equations with Euclidean symmetry. *Arch. Rational Mech. Anal.* **131** (1995) 199–224.

[2] P.C. Bressloff, J.D. Cowan, M. Golubitsky, and P.J. Thomas. Scalar and pseudoscalar bifurcations motivated by pattern formation on the visual cortex, *Nonlinearity* **14** (2001) 739–775.

[3] D.R.J. Chillingworth, E. Vicente Alonso, and A.A. Wheeler. Geometry and dynamics of a nematic liquid crystal in a uniform shear flow. *J. Phys. A* **34** (2001) 1393–1404.

[4] P. Chossat and R. Lauterbach. *Methods in Equivariant Bifurcations and Dynamical Systems*, World Scientific, Singapore 2000.

[5] P.G. de Gennes. *Mol. Cryst. Liq. Cryst.* **12** (1971) 193.

[6] P.G. de Gennes. *The Physics of Liquid Crystals*, Clarendon Press, Oxford 1974.

[7] M. Golubitsky and I. Stewart. *The Symmetry Perspective: From Equilibrium to Chaos in Phase Space and Physical Space*, Birkhäuser, Basel, 2002.

[8] M. Golubitsky, I. Stewart, and D.G. Schaeffer. *Singularites and Groups in Bifurcation Theory*, II, Springer-Verlag, New York 1988.

[9] H. Grebel, R.M. Hornreich, and S. Shtrikman. *Phys. Rev.* A **28** (1983) 1114–1138.

[10] J.-H. Huh, Y. Hidaka, A.G. Rossberg and S. Kai. *Phys. Rev.* E **61** (2000) 2769.

[11] J.-H. Huh, Y. Hidaka, Y. Yusuf, S. Kai, N. Éber and Á. Buka. Abstract 24A-4-1 ILCC 2000 (Sendai).

[12] J.-H. Huh, Y. Hidaka and S. Kai. Abstract 24D-87-P ILCC 2000 (Sendai).

[13] S. Kai and K. Hirakawa. *Solid State Comm.* **18** (1976) 1573.

[14] P.D. Olmsted and P.M. Goldbart. *Phys. Rev.* A **46** (1992) 4966.

[15] T.J. Sluckin. The liquid crystal phases: physics and technology, *Contemporary Physics* **41** (2000) 37–56.

[16] E. Vicente Alonso, PhD thesis, University of Southampton, 2000.

Martin Golubitsky
Department of Mathematics
University of Houston
Houston, TX 77204-3476, USA
e-mail: mg@uh.edu

David Chillingworth
Department of Mathematics
University of Southampton
Southampton SO17 1BJ, UK
e-mail: drjc@maths.soton.ac.uk

Trends in Mathematics:
Bifurcations, Symmetry and Patterns, 67–74
© 2003 Birkhäuser Verlag Basel/Switzerland

Patchwork Patterns:
Dynamics on Unbounded Domains

Peter Ashwin

Abstract. We discuss some problems concerning the asymptotic behaviour of patterns generated by evolution equations on unbounded domains. We suggest an approach using a number of different topologies to examine the asymptotic behaviour of patterns. This highlights some problems that need to be understood in constructing a topological theory of dynamics for spatio-temporal patterns.

1. Introduction

The spatio-temporal dynamics of evolution equations on unbounded (infinite) domains is notoriously difficult to analyse, especially with regard to understanding the asymptotic dynamics or attractors. This is often due to subtleties of the functional analytic setting that is required to give an adequate setting for these problems. However there is also a more fundamental problem that is also the root of many of the former difficulties; in the simplest terms this is the existence of two quantities that are unbounded, namely space and time. At best this gives rise to a tradeoff between two limiting processes; at worst it gives rise to lack of compactness or completeness that mean that in many topologies the ω-limit sets are empty.

In the last ten years there has been considerable progress towards understanding the qualitative dynamics for specific PDEs on unbounded domains, see for example [5, 6, 10, 7, 12, 11] who use a variety of topologies on the space of unbounded patterns to obtain significant results on the existence, attraction and stability properties of global attractors.

For example, in their work [10] Mielke and Schneider work with a function space consisting of the subspace of functions in the Sobolev space and a weighted norm such that the weighted norm remains finite under translations and the action of translation is continuous. Using this they are able to prove many properties of a global attractor for the Ginzburg-Landau equation in this setting and relate this to properties of the Swift-Hohenberg equation. In particular they obtain a compact attractor in the weighted norm, though it is not compact in any translation-invariant norm.

At the same time there has been a realisation that exploiting noncompact group structure that is often present in such systems can say a lot about the problem independent of the particular classes of equations. This has been particularly successful in describing many properties of localised patterns such as spiral waves [3, 12, 2]. In this paper we examine some issues in trying to combine these approaches to give a topological description of continuous evolutions of patterns. Rather than examining specific models using properties of spectra we do not assume any differentiability and aim for a purely topological description on the premise that only when this is understood can one expect to get a good qualitative understanding of PDE or other spatially extended model equations.

2. Topologies for patterns

We consider $\mathbb{R}^r$-valued patterns on some space of patterns on some $\mathbb{R}^d$ ($d, r \geq 1$) with a Lie group Γ that acts transitively on $\mathbb{R}^d$. Such a group must necessarily be noncompact; typically we consider Γ is the Euclidean group $\mathbf{E}(d)$ or some subgroup of this. Most of the ideas below can of course be easily extended to patterns on domains that are products of $\mathbb{R}^d$ with any compact connected space. The connected component of Γ we write as Γ_0.

We write $\rho(y)$ to mean the action of $\rho \in \Gamma$ on $y \in \mathbb{R}^d$ and note that this action extends naturally to functions $u : \mathbb{R}^d \to \mathbb{R}^r$ in the usual way by action on the domain:

$$\rho(u)(x) = u(\rho^{-1}x).$$

Let $\mathcal{C}^d$ be the space of all compact subsets of $\mathbb{R}^d$ equipped with the usual Hausdorff metric $h(A, B) = \sup_{a \in A} \inf_{b \in B} |a - b| + \sup_{b \in B} \inf_{a \in A} |a - b|$. Let $I_C(x)$ be the indicator function such that $I_C(x) = 1$ if $x \in C$ and $I_C(x) = 0$ otherwise. Consider the function space

$$X \subset C^0_{ub}(\mathbb{R}^d, \mathbb{R}^r)$$

of bounded uniformly continuous functions with a global norm (for example $\|u\|_g = \sup_x |u(x)|$) and a local norm (for example $\|u\|_l = \int |u(x)|^2 \, dx$); these define a *global* and a *local topology* on X. We assume that:

1. X is complete with respect to $\|\cdot\|_g$.
2. For any $C \in \mathcal{C}^d$ and $u \in X$ we require that $\|uI_C\|_l < \infty$ (though $\|u\|_l$ itself may be infinite).
3. Γ acts continuously on X and the norms $\|\cdot\|_{g,l}$ are invariant under Γ, ie $\|\rho u\|_{g,l} = \|u\|_{g,l}$ for all $u \in X$ and $\rho \in \Gamma$.

Under these assumptions we say $u \in X$ is a *pattern* and X is a *pattern space*.

2.1. Semiflows of patterns

Suppose that $\Phi : X \times \mathbb{R}^+ \to X$ is a semigroup on a pattern space X, i.e., a map such that

$$\Phi_0 u = u$$
$$\Phi_s \circ \Phi_t u = \Phi_{s+t} u$$

holds for any $u \in X$ and $s, t \geq 0$. We require that $\Phi_t(u)$ is continuous in $t \in \mathbb{R}^+$ and $u \in X$ w.r.t. the global topology on X. We suppose that Φ_t is equivariant under the action of Γ on X, i.e.,

$$(\rho \circ \Phi_t)u = (\Phi_t \circ \rho)u$$

for any $t > 0$, $\rho \in \Gamma$ and $u \in X$. As usual, given a pattern $u \in X$ we define the *isotropy* of u to be

$$\Sigma(u) = \{\gamma \in \Gamma \; : \; \gamma u = u\}.$$

This is a subgroup of Γ that expresses its symmetry; for example 'roll-like' solutions of a system with $\mathbf{E}(2)$ symmetry on a planar unbounded domain have isotropy $\mathbf{E}(1)$.

2.2. The patch topology

The assumptions above ensure that there is another topology on the pattern spaces X, namely the weakest topology where

$$u(x, t) \to_p v(x)$$

if and only if

$$\|(u(x, t) - v(x))I_C\|_l \to 0 \text{ as } t \to \infty$$

for all $C \in \mathcal{C}^d$. In other words, a set U is a neighbourhood of $v \in X$ in this topology if and only if there exists a positive function $\psi : \mathcal{C}^d \to \mathbb{R}^+$ such that

$$\|(u(x) - v(x))I_C\|_l < \psi(C)$$

implies that $u \in U$. We refer to this as the *patch topology* $\mathcal{T}_p$.

Recall that a topology is *Hausdorff* if given any $u \neq v$ there exist disjoint neighbourhoods U of u and V of v; see for example [8, 13]. It is clear that the patch topology is Hausdorff; given any two $u \neq v$ we can find separating neighbourhoods by virtue of the assumption that they are subsets in C_{ub}^0.

2.3. The patch orbit topology

We now define a topology based on closeness over group orbits: the *patch orbit topology* $\mathcal{T}_{po}$. This is the weakest topology where

$$u(x, t) \to_{po} v(x)$$

if and only if

$$\inf_{\rho \in \Gamma} \|(\rho u(x, t) - v(x))I_C\|_l \to 0 \text{ as } t \to \infty$$

for all $C \in \mathcal{C}^d$. This topology is natural in the sense that in this topology, a pattern u is close to another pattern v if some translate of u matches v on patches of arbitrarily large area.

Lemma 2.1. *For any $d > 0$ the topological space $(X, \mathcal{T}_{po})$ is not Hausdorff.*

Proof. We prove for $d = 1$; this can be extended easily to any dimension. Take $u_1 = \sin x$ and $u_2 = \sin 2x$. Consider a function

$$u(x) = \sin\left(x\frac{2 + e^x}{1 + e^x}\right)$$

or any similar function that behaves like $\sin x$ for large $x > 0$ and like $\sin 2x$ for large $x < 0$. Note that any neighbourhood of u contains both u_1 and u_2. It follows (eg [8, p88]) that the topology is not T_1 and hence $(X, \mathcal{T}_{po})$ is not Hausdorff. $\square$

Note that although $(X, \mathcal{T}_{po})$ is not Hausdorff the quotient obtained by identifying points on the same group orbit may be Hausdorff.

2.4. Propagating Patch Topology

The final topology we define sits between the patch and the patch orbit topologies. The *propagating patch* topology $\mathcal{T}_{pp}$ is the weakest topology such that

$$u(x, t) \to_{pp} v(x)$$

if and only if there is a continuous function $\rho : \mathbb{R}^+ \to \Gamma$ with $\rho(0) = \text{Id}$ such that

$$\|(\rho(t)u(x, t) - v(x))I_C\|_l \to 0 \text{ as } t \to \infty$$

for all $C \in \mathcal{C}^d$. We refer to the function $\rho(t)$ as the *patch path* that gives convergence of u to v. For any $C \in \mathcal{C}^d$ we say $\rho^{-1}(t)C$ is a *propagating patch*; this topology measures two patterns as close if they approach each other on all propagating patches of this form.

In the case that Γ is not connected then ρ must remain within the connected component Γ_0, for example $\mathbf{SE}(2)$ instead of $\mathbf{E}(2)$. We write the set of propagating patch limits as

$$\lim_{pp} u(x, t) = \{v \in X \ : \ u(x, t) \to_{pp} v(x) \text{ as } t \to \infty\}$$

and note that even if $u(x, t)$ is time independent then its propagating patch limits may include more than just its group orbit under Γ.

Theorem 2.2. *For any $u, v \in X$ we have that $u \to_p v$ implies that $u \to_{pp} v$ implies that $u \to_{po} v$.*

Proof. Note that for any $C \in \mathcal{C}^d$ and $u, v \in X$ we have

$$\|(u - v)I_C\|_l \geq \inf_{\rho \in \Gamma_0} \|(\rho u - v)I_C\|_l \geq \inf_{\rho \in \Gamma} \|(\rho u - v)I_C\|_l$$

and the result follows. $\square$

By adapting the examples in Section 3, one can demonstrate that implications of this theorem cannot be reversed. If we can choose the patch path ρ so that

$$\|(\rho(t)u(x, t) - v(x))I_C\|_l$$

decays exponentially fast then all its time derivatives will also decay. This suggests that if u approaches a relative equilibrium v such that $\Sigma(v)$ is compact then the patch path will converge to a one parameter group in Γ_0, and this in turn may

Pattern	(a)	(b)	(c)	(d)	(e)
Isotropy	Id	$E(1)$	$\mathbb{Z} \times E(1)$	$E(1)$	$E(2)$

TABLE 1.   Isotropy types of the limiting patterns depicted in Figure 2.

provide a link to the generic drifts studied in [1]. In fact, the question of bounded or unbounded drifts of relative equilibria can be reinterpreted in the following way:

Proposition 2.3. *Suppose that $u(x,t) = \exp(\xi t)v(x)$ is an evolution of a relative equilibrium with $\xi \in L\Gamma_0$. Then $u \to_{pp} v$. Moreover, $u \to_p v$ if and only if the drift is bounded, i.e., if and only if $\overline{\{\exp \xi t \ : \ t \in \mathbb{R}\}}$ is a compact subgroup of Γ_0.*

3. Case studies

3.1. Convergence to a spiral wave

A technique that is used both experimentally and numerically to construct planar spiral waves for reaction-diffusion systems that are excitable about a homogeneous steady state is as follows (see for example [9, 14, 4]). We use this example to illustrate the use of the propagating patch limit.

1. Take a single infinite wavefront that is localised in x_1, propagating towards positive x_1, and parallel to the x_2 direction.
2. The wavefront in the half-plane $x_2 > 0$ is replaced by the homogeneous state. This creates a semi-infinite wave with a discontinuous tip; the tip is immediately smoothed by diffusion.
3. At certain parameter values the tip describes a circular motion and the wavefront winds up to form a spiral.

Figure 1(i-iv) schematically shows the formation of a spiral by this 'kinetic' mechanism. If we use the global topology C_{ub}^0 on X, the pattern never approaches the spiral state; in fact its ω-limit is empty! The only relative equilibrium that is approached by this pattern in the *patch topology* is the group orbit of an infinitely wound spiral shown in Figure 2(a); however in the *propagating patch topology* there are limits shown in Figure 2(a-e). This can be seen by taking patches that propagate with centres at the points marked (a-e) on Figure 1.

Note that multiple limits in the propagating patch topology are possible. Moreover the limits can have a variety of isotropy types; in Figure 2 we see that the set $\lim_{pp} u(x,t)$ includes relative equilibria with a variety of isotropies listed in Table 1. Observe that some isotropies are compact and others noncompact. Each of these patterns propagates generically according to its symmetry.

3.2. Relaxation of a defect in one dimension

The second example we examine is the heat equation $u_t = u_{xx}$ in one dimension (see [6, p1294]). This has solutions of the form

$$u(x,t) = \lambda_- + (\lambda_+ - \lambda_-)\mathrm{erf}(x/(2\sqrt{t}))$$

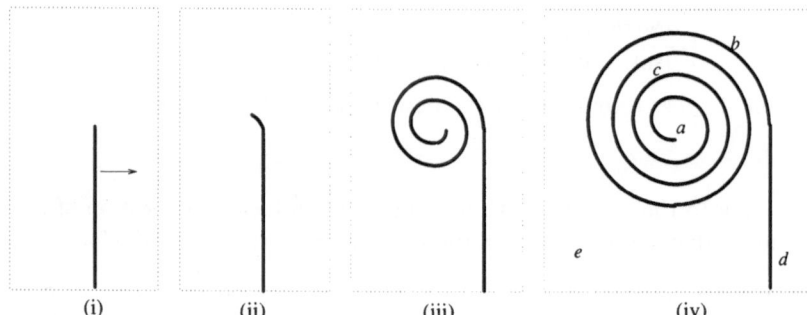

FIGURE 1. Diagram showing formation of an infinite spiral wave from propagation of a semi-infinite wave as time progresses from (i) to (iv). The locations a-e on (iv) refer to centres of propagating patches that can be taken to get convergence to the patch path limits shown in Figure 2. The arrows show the sense of motion of the wavefronts and the squares the location of the centre of rotation.

which are asymptotic to $\lambda_\pm$ as $x \to \pm\infty$. Given any $\lambda \in [\lambda_-, \lambda_+]$ we can find a propagating patch such that $u(x,t) \to_{pp} \lambda$ where λ is the constant pattern $v(x) = \lambda$. Thus the propagating patch limit may contain a continuum of patterns. This example can clearly be adapted to give convergence to a continuum of periodic patterns.

3.3. Absolute and convective instabilities

The fixed and propagating patch topologies suggest possible nonlinear analogies for absolute and convective linear stability respectively (see eg [5]).

We say a pattern $v(x)$ is *fixed patch stable* w.r.t. u if the evolution $u(x,t)$ is such that $u \to_p v$ and given any evolution $w(x,t)$ with $\|u(x,0) - w(x,0)\|_g$ small enough we also have $w \to_p v$. Similarly we say $v(x)$ is *propagating patch stable* w.r.t. $u(x,t)$ if all evolutions $w(x,t)$ with sufficiently small $\|u(x,0) - w(x,0)\|_g$ have $w \to_{pp} v$. A long term aim would be to be able to use absolute/convective stability properties of spectra (e.g., [5, 11]) to prove existence of fixed/propagating patch stable patterns.

4. Discussion

We suggest that there is a need to develop a 'pre-differentiable' theory for describing continuous evolutions of patterns. In doing so we have come up with some ideas of how one can describe attraction and (convective) stability of patterns without reference to spectral properties of governing equations. Using this we see there may be many unbounded patterns that can be reasonably found in a single parametrized family of patterns; these patterns can have a range of symmetries and can translate and rotate differently in different parts of the domain.

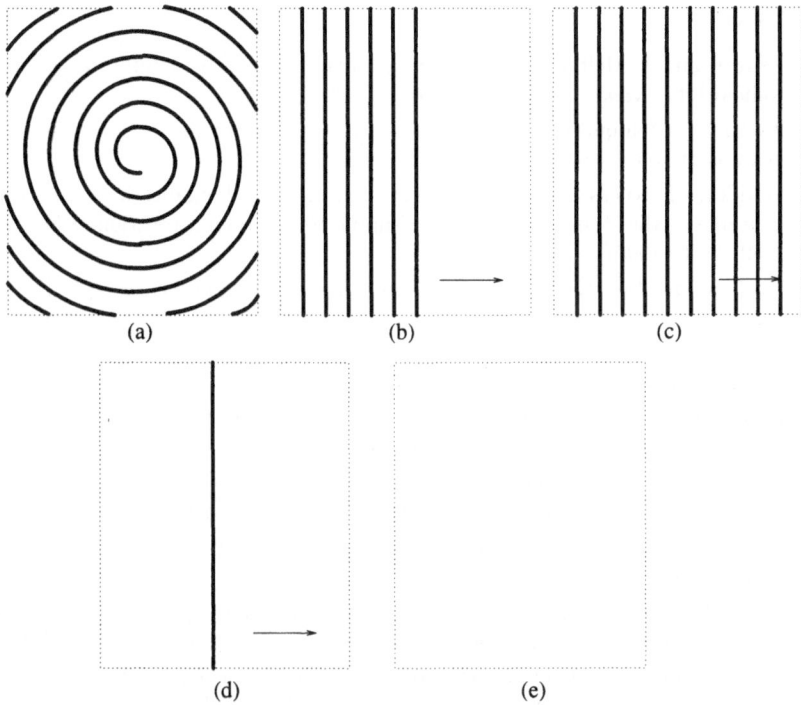

FIGURE 2. The sequence of patterns shown in Figure 1 limits in the
propagating patch topology to the pattern shown in this figure. The
spiral shown in (a) is the unique limit in the patch topology. The group
orbit (under $SE(2)$) of each of these patterns is contained in the set of
propagating patch limits.

We do not say anything about the 'fronts' that may exist between patterns; this
remains an interesting challenge. One intriguing question is whether these ideas
can be usefully developed to understand evolutions on large domains where finite patches are observed; for example in weak spiral turbulence of the complex
Ginzburg-Landau equation there are a number of spirals that jostle against each
other. Given a sensible notion of convergence we can start to discuss attraction,
stability and bifurcation of patterns on infinite domains, but until we have this
it remains a very difficult problem (see for example the transition from spiral to
retracting wave [4, 14, 2]). Linearised stability on unbounded domains is fraught
with problems associated with the unboundedness; linearised eigenfunctions are
often unbounded (and so are not in C_{ub}^0) and we expect that they will describe
behaviour on bounded regions.

Acknowledgements

The author thanks Ian Melbourne for stimulating discussions concerning this work.

References

[1] P. Ashwin and I. Melbourne. Noncompact drift for relative equilibria and relative periodic orbits. *Nonlinearity* **10**:595-616 (1997).

[2] P. Ashwin, I. Melbourne and M. Nicol. Drift bifurcations of relative equilibria and transitions of spiral waves. *Nonlinearity* **12**:741–755 (1999).

[3] D. Barkley, M. Kness and L. Tuckerman. Spiral-wave dynamics in a simple model of excitable media: the transition from simple to compound rotation. *Phys. Rev. A* **42**:2489–2491 (1990).

[4] D. Barkley and I.G. Kevrekidis A dynamical systems approach to spiral wave dynamics. *Chaos* **4**:1-8 (1994).

[5] P. Collet and J.-P. Eckmann. *Instabilities and Fronts in Extended Systems*. Princeton University Press (1990).

[6] P. Collet and J.-P. Eckmann. Space-time behaviour in problems of hydrodynamic type: a case study. *Nonlinearity* **5**:1265–1302 (1992).

[7] E. Feireisl, P. Laurençot and F. Simondon. Global attractors for degenerate parabolic equations on unbounded domains. *J. Diff. Eqns.* **129**:239–261 (1996).

[8] I.M. James. *Topologies and Uniformities*, Springer Undergraduate Mathematics Series, Springer-Verlag London (1999).

[9] A. S. Mikhailov and V. S. Zykov. Kinematical theory of spiral waves in excitable media: Comparison with numerical simulations. *Physica D* **52**:379-397 (1991).

[10] A. Mielke and G. Schneider. Attractors for modulation equations on unbounded domains- existence and comparison. *Nonlinearity* **8**:743–768 (1995).

[11] B. Sandstede and A. Scheel. Essential instabilities of fronts: bifurcation and bifurcation failure. *Dynamics and Stability of Systems* **16**:1–28 (2000).

[12] B. Sandstede, A. Scheel and C. Wulff, Bifurcations and dynamics of spiral waves *J. Nonlinear Sci.* **9**:439–478 (1999).

[13] B.T. Sims. *Fundamentals of Topology* Macmillan, New York (1976).

[14] W. Jahnke and A. T. Winfree. A survey of spiral-wave behaviors in the Oregonator model. *Int. J. Bifurcation. Chaos* **1**:445–466 (1991).

Peter Ashwin
School of Mathematical Sciences
University of Exeter
Exeter EX4 4QE, UK.
e-mail: P.Ashwin@ex.ac.uk

Trends in Mathematics:
Bifurcations, Symmetry and Patterns, 75–86
© 2003 Birkhäuser Verlag Basel/Switzerland

Persistent Ergodicity and Stably Ergodic SRB Attractors in Equivariant Dynamics

Michael Field

Abstract. We describe some recent analytic results on the co-existence of symmetry and chaotic dynamics in equivariant dynamics. We emphasize the case of skew-products and stably SRB attractors.

1. Introduction & Background

It has been known for some time that the assumption of symmetry can lead to robust complex dynamics in low dimensional systems. At the same time, the presence of symmetry can also lead to regularity. It was observed in 1988 by Chossat & Golubitsky [10], that both complexity and regularity can co-exist in the dynamics of symmetric polynomial mappings of the plane. In this case, the regularity appears 'on average' and is best expressed in terms of the existence of a symmetric (physical) measure. In Figure 1 we show a characteristic example of a numerically computed attractor for a non-invertible planar polynomial map with $\mathbf{D}_4$ symmetry.

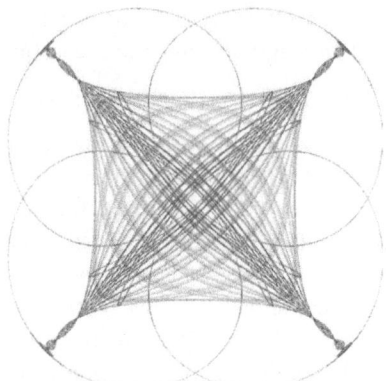

FIGURE 1. Attractor of a planar map with $\mathbf{D}_4$ symmetry.

Received by the editors October 10, 2001
Research supported in part by NSF Grant DMS-0071735.

Although the image shown in Figure 1 appears to be $\mathbf{D}_4$ symmetric, the symmetry is only approximate. Exact symmetry can be obtained through a limiting process and is best expressed in terms of the existence of a symmetric *physical* or *Sinai-Ruelle-Bowen* (SRB) measure on the attractor (see §1.1).

Although it is rather easy to find numerical examples of 'symmetric chaos' in planar polynomial maps, it is notoriously difficult to establish rigorous analytic results that prove the existence of chaotic dynamics and physical measures. Indeed, just as for one dimensional maps of the line, small changes in the map can lead to 'windows' of attracting periodic points.

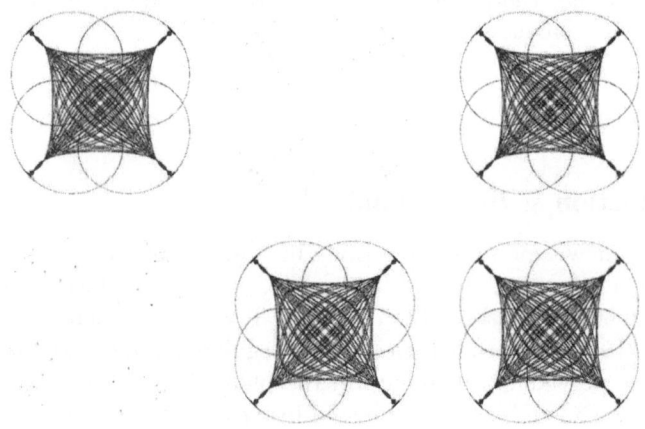

FIGURE 2. The effect of slowly varying parameters

In Figure 2 we show the effect of slowing varying a parameter in the map that generated Figure 1. The original map was given by

$$f(z) = (1.52 - 1.39|z|^2 - 0.0901\mathrm{Re}(z^5))z - 0.8005\bar{z}^3.$$

The sequence shown in Figure 2 was obtained by successively incrementing the coefficient of $|z|^2$ by -0.00014. Reading from the top left, the second image is a periodic orbit (with $\mathbf{Z}_5$ symmetry). The fourth image may look like a periodic orbit but it is a 'Henon-like' attractor. In Figure 3 we show the magnification[1] of a very small region from Figure 3 which lies inside one of the 'dots'.

The 'random' transitions between chaotic attractors and attracting periodic orbits shown in Figure 2 suggests that it is likely to be very difficult, if not impossible, to find computable conditions on the generating map that imply chaotic behavior. However, if we want to build a theory of chaotic dynamics for equivariant maps, it is vital to have a library of good examples where we can analytically prove the presence of chaotic dynamics. To this end, for most of the remainder of

---

[1]The magnification is by a factor of 21,205 in both the x- and y-directions.

FIGURE 3. Magnification of a small region in Figure 2, image 4

this article, we shall be assuming that maps are smooth and invertible and that attractors possess at least some uniform hyperbolic structure.

1.1. Attractors

It is time to be more precise about exactly what we mean by chaotic dynamics on an attractor.

Definition 1.1. *Let* $f : \mathbb{R}^n \to \mathbb{R}^n$ *be smooth*[2]*. A compact f-invariant subset A of $\mathbb{R}^n$ is an* attractor *if*

(a) *There exists an open neighborhood U of A such that $f(\bar{U}) \subset U$ and $\cap_{n \geq 0} f^n(U) = A$.*

(b) $f : A \to A$ *is transitive.*

Definition 1.2. *An attractor A is* SRB *('Sinai-Ruelle-Bowen'), if there exists an f-invariant ergodic measure μ on A, such that for Lebesgue almost all $x \in U$,*

$$\lim_{n \to \infty} \frac{1}{n} \sum_{i=0}^{n-1} \delta_{f^i(x)} = \mu.$$

The measure μ, necessarily unique, is called the SRB *measure (on A).*

Remark 1.3. The reader is cautioned that there are several different definitions of SRB measure commonly used in the literature [28]. Under some of these definitions there may be more than one SRB measure on an attractor. However, for Axiom A attractors, all these definitions are equivalent [4]. If we do not assume Axiom A and there are zero Liapunov exponents, then, without further conditions, our definition does not imply the measure has absolutely continuous conditional measures on unstable manifolds or that the measure is an equilibrium state for the Jacobian potential.   ◇

If A is an SRB attractor for f, we say A is *robust* if, for all g sufficiently C^1-close to f, there exists a compact g-invariant set $A(g)$, close to A, which is an SRB attractor for g. We also say A is *stably SRB*.

In the following, we always assume an SRB attractor is connected and consists of more than one point. In particular, A will not be a fixed or periodic point and there will be complex dynamics on A.

Let Γ be a compact Lie group and $(\mathbb{R}^n, \Gamma)$ be a Γ-representation. A smooth map $f : \mathbb{R}^n \to \mathbb{R}^n$ is Γ-equivariant if $f(\gamma x) = \gamma f(x)$, for all $x \in \mathbb{R}^n$, $\gamma \in \Gamma$.

---

[2]That is, smooth enough. We usually assume C^∞, though C^1 or C^2 will often suffice.

Suppose that A is an attractor for the Γ-equivariant map $f : \mathbb{R}^n \to \mathbb{R}^n$. We define the symmetry group Σ_A of A by

$$\Sigma_A = \{\gamma \in \Gamma \mid \gamma A = A\}.$$

Using our definition of attractor, it is easy to show that γA is an attractor for all $\gamma \in \Gamma$ and that $\Sigma_{\gamma A} = \gamma \Sigma_A \gamma^{-1}$. Moreover, if $\gamma \in \Gamma \setminus \Sigma_A$, then $A \cap \gamma A = \emptyset$. See [10] for a proof and applications.

In the remainder of this article, we shall describe recent results about the existence of robust SRB attractors for Γ-equivariant diffeomorphisms. We start by looking at the case where Γ is a finite group and then proceed to examine the case where Γ is a connected (non-finite) compact Lie group.

2. Attractors for systems equivariant by a finite group

In the case when there is no symmetry, there is a well-developed theory of hyperbolic invariant sets for diffeomorphisms (see [19]). In particular, if an attractor has a (uniform) hyperbolic structure, then it has a unique SRB measure.

If Γ is finite, then there are no dimensional obstructions to hyperbolicity as Γ-orbits are zero dimensional. Hence, it is natural to start by restricting to the class of connected *hyperbolic* H-invariant attractors of Γ-equivariant diffeomorphisms, where H is a subgroup of Γ. In this case (see above), hyperbolicity implies that every H-invariant attractor admits a unique SRB measure. Further, the uniqueness implies that the measure is H-invariant. Referring to Figure 1, the conjectured SRB measure is $\mathbf{D}_4$-invariant. (However, the map used to generate Figure 1 is not invertible, the attractor is not hyperbolic and presently there are no available techniques to prove the existence of an SRB measure on the attractor.)

If we assume that the symmetry group of the attractor acts freely on the attractor, then it is possible to give necessary and sufficient conditions on the representation $(\mathbb{R}^n, \Gamma)$ for there to exist an equivariant diffeomorphism $f : \mathbb{R}^n \to \mathbb{R}^n$ which has a connected hyperbolic SRB attractor with symmetry group equal to H[3]. Hyperbolic attractors with specified symmetry group may be constructed by combining Williams' theory of expanding attractors [31] with symmetry arguments. We refer the reader to [14] for proofs and statements of the general results (which also apply to flows). We give a special case of one general theorem proved in [14] as well as an example that illustrates some of the main techniques.

Theorem 2.1 ([14, Theorem 1.4]). *Suppose that $n \geq 4$ and that Γ contains no reflections. Then for any subgroup H of Γ, we may find a smooth Γ-equivariant diffeomorphism of $\mathbb{R}^n$ which has a hyperbolic attractor A with $\Sigma_A = H$. Further, we may require that H acts freely on A.*

Example 2.2. We take the action of $\mathbb{Z}_2$ on $\mathbb{R}^3$ defined by $(x, y, z) \mapsto (-x, -y, z)$. This action has fixed point set $\mathrm{Fix}(\mathbb{Z}_2)$ equal to the z-axis. We briefly sketch the

---

[3]For results on non-invertible maps, see[3].

construction of a $\mathbb{Z}_2$-equivariant diffeomorphism of $\mathbb{R}^3$ which has a $\mathbb{Z}_2$-symmetric hyperbolic attractor. The construction goes in three stages. First we choose a smooth $\mathbb{Z}_2$-equivariant expanding map f of a smooth $\mathbb{Z}_2$-invariant graph $G \subset \mathbb{R}^3$ that satisfies Williams' conditions at the vertices of the graph. Let T be a $\mathbb{Z}_2$-invariant tubular neighborhood of G. Following Williams [31], we realize the inverse limit of $f : G {\to} G$ as a solenoidal attractor of a smooth $\mathbb{Z}_2$-equivariant embedding $F : T {\to} T$. Finally, we must show that for our choice of f, F is smoothly equivariantly isotopic to the identity mapping of T. It then follows by the equivariant isotopy theorem that F extends smoothly to an equivariant diffeomorphism of $\mathbb{R}^3$. We have to be careful with the last step. For example, if we take G to be the unit circle in the x,y-plane and $f(\theta) = 3\theta$, then we cannot *equivariantly* isotop the resulting $F : T {\to} T$ to the identity map on T (the z-axis blocks any equivariant isotopy).

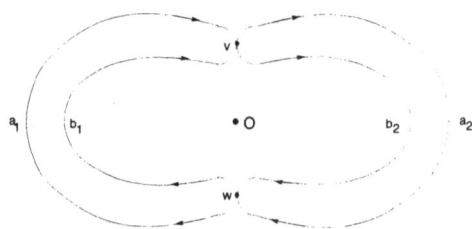

FIGURE 4. A smooth $\mathbb{Z}_2$-invariant graph

In Figure 4, we show a smooth $\mathbb{Z}_2$-invariant graph G with vertices v, w and (directed) edges a_1, b_1, a_2, b_2. We regard the graph as $\mathbb{Z}_2$-equivariantly embedded in the x,y-plane. We define a smooth expanding map $f : G {\to} G$ according to the $\mathbb{Z}_2$-symmetric set of edge rules

$$f(a_1) \;=\; a_2^{-1}b_2a_1, \quad f(b_1) \;=\; b_2^{-1}a_2b_1,$$
$$f(a_2) \;=\; a_1^{-1}b_1a_2, \quad f(b_2) \;=\; b_1^{-1}a_1b_2.$$

Since the edge rules are $\mathbb{Z}_2$-symmetric, it is easy to see that we can require f to be $\mathbb{Z}_2$-equivariant. Finally, observe that the image of each edge E by f is an arc $A(E)$ joining the two vertices v, w and that $A(E)$ can be equivariantly isotoped to E within $\mathbb{R}^2$. Using this it is not hard to see that we may require the corresponding map $F : T {\to} T$ to be equivariantly isotopic to the identity on T. We refer to [14, §§3,4] for details.   ♡

Our constructions provide a large class of stably SRB symmetric attractors. Indeed, every hyperbolic SRB attractor is stably SRB since hyperbolicity is an open condition for attractors and connected hyperbolic SRB attractors admit a unique SRB measure (of course, the SRB measure may change when we perturb the map).

Matters are very different if Γ is not finite and Γ acts freely on an attractor A. In this case, the attractor can never be hyperbolic.

3. Skew extensions by compact connected Lie groups

Let $\Phi : N \to N$ be a smooth diffeomorphism of the compact manifold N. Suppose that $\Lambda \subset N$ is a hyperbolic attractor for Φ. Set $\phi = \Phi|\Lambda$. We say that a map defined on Λ is smooth if it extends to a smooth map defined on some open neighborhood of Λ (in N).

We assume that Γ is a compact and connected Lie group. Let $C^\infty(\Lambda, \Gamma)$ denote the space of smooth *cocycles* $f : \Lambda \to \Gamma$. For $\infty \geq r \geq 0$, we define the corresponding C^r topology on $C^\infty(\Lambda, \Gamma)$. If $f \in C^\infty(\Lambda, \Gamma)$, define $\phi_f : \Gamma \times \Lambda \to \Gamma \times \Lambda$ by

$$\phi_f(\gamma, \lambda) = (\gamma f(\lambda), \phi(\lambda)), \quad (\gamma \in \Gamma, \lambda \in \Lambda).$$

We have a natural *left* action of Γ on $\Gamma \times \Lambda$ and, with respect to this action, ϕ_f is Γ-equivariant. We say that ϕ_f is a Γ-*extension* or *skew extension* of ϕ by f.

Let μ denote the SRB measure on Λ and h denote *Haar* measure on Γ. Set $\nu = h \times \mu$. Then, for all cocycles f, ϕ_f is a ν-measure preserving diffeomorphism of $\Gamma \times \Lambda$.

Basic questions: Can we choose f so that $\Gamma \times \Lambda$ is an SRB attractor with SRB measure ν? Can $\Gamma \times \Lambda$ be a robust SRB attractor for ϕ_f (within the class of Γ-equivariant diffeomorphisms of $\Gamma \times N$)?

3.1. An example: The skew cat map

Let λ denote the area measure on the 2-torus $\mathbb{T}^2 = \mathbb{R}^2/\mathbb{Z}^2$ induced from Lebesgue measure on $\mathbb{R}^2$. Take the Thom-Anosov diffeomorphism (cat map) $\hat{A} : \mathbb{T}^2 \to \mathbb{T}^2$ defined by the matrix $A = \begin{pmatrix} 2 & 1 \\ 1 & 1 \end{pmatrix}$. Since $\det(A) = 1$, $\hat{A}$ is a λ-measure preserving diffeomorphism of $\mathbb{T}^2$. It is well known that $\hat{A}$ is λ-ergodic [29] and hence λ is SRB (λ is already Lebesgue). Let $c \in SO(2)$ and define the $SO(2)$-extension

$$\hat{A}_c : SO(2) \times \mathbb{T}^2 \to SO(2) \times \mathbb{T}^2,$$

by $\hat{A}_c(\theta, t) = (\theta + c, \hat{A}(t))$. Although $\hat{A}_c$ is never hyperbolic, it is hyperbolic *transverse* to the action of $SO(2)$. That is, transverse to the group orbits. The map $\hat{A}_c$ is an example of a *partially hyperbolic* diffeomorphism [7, 25].

It is easy to verify that $\hat{A}_c$ is ergodic if and only if c defines an irrational rotation. Of course, $\hat{A}_c$ can never be stably SRB. Nevertheless, there is a C^0-open, C^∞ dense set of smooth cocycles f for which $\hat{A}_f$ is stably SRB. With slightly different hypotheses and conclusions, this result follows from a general theorem of Brin [5, 6]. Much later, Adler, Kitchens, Shub [1] reproved the result and obtained an elegant characterization of stably SRB related to the Livšic periodic point theorem. Specifically, they proved that $\hat{A}_f$ is stably SRB if and only if there exists a pair of periodic points p, q of $\hat{A}$ with the same period, N say, such that

$$\Pi_{i=0}^{N-1} f(\hat{A}^i(p)) \neq \Pi_{i=0}^{N-1} f(\hat{A}^i(q)).$$

The C^0-openness, C^∞ density result follows immediately from this inequality. Later, Burns & Wilkinson [8] showed[4] that if $\hat{A}_f$ is stably SRB within the class of skew extensions, then it is stably SRB within the class of smooth volume preserving diffeomorphisms of $\mathrm{SO}(2) \times \mathbb{T}^2$.

3.2. Stable Ergodicity

Before describing some recent results on robust ergodicity for attractors of equivariant maps, we briefly review some of the recent developments on stable ergodicity both for symmetric and for volume preserving diffeomorphisms (no symmetry).

A (volume preserving) diffeomorphism of a compact manifold is called *stably ergodic* if it is ergodic and any small volume preserving perturbation of it remains ergodic.

A basic question is to understand the typicality of ergodicity for smooth dynamical systems. In 1967 Anosov [2] proved that hyperbolic diffeomorphisms preserving a measure equivalent to a volume are ergodic. Since hyperbolicity is a $(C^1$-) open condition, Anosov systems give examples of open sets of stably ergodic transformations. Unfortunately, hyperbolicity is a very strong condition. In 1994, Grayson, Pugh & Shub [18] showed that the time-1 map of the geodesic flow on a surface of constant negative curvature was stably ergodic (within the class of volume preserving diffeomorphisms). The time-1 map is not hyperbolic but is partially hyperbolic (hyperbolic transverse to trajectories of the flow). Later, Pugh and Shub [25] conjectured that among *partially* hyperbolic volume preserving C^2-diffeomorphisms stable ergodicity is dense (and automatically open). Recently, many new results have been proved on stable ergodicity for volume preserving diffeomorphisms. We refer to [9] for a recent survey.

The first results on stable ergodicity within the class of skew extensions by compact Lie groups were obtained by Brin in 1975. Brin, in the context of frame flows on negatively curved manifolds, proved that skew extensions over an Anosov system were generically C^1-stably ergodic [6]. Using results of Parry [23], Parry and Pollicott [24] extended Brin's results to a large class of toral extensions (connected compact abelian Lie groups). Specifically, they proved the following

Theorem 3.1. *Let Λ be a hyperbolic basic (locally maximal) set. Assume that* either Λ *is a subshift of finite type or* Λ *is connected and* $\phi^\star : H^1(\Lambda, \mathbb{Z}) \to H^1(\Lambda, \mathbb{Z})$ *does not have one as an eigenvalue. Then there is an open (C^α-topology, $\alpha > 0$) and dense (C^∞-topology) subset $\mathcal{U}$ of $C^\infty(\Lambda, \mathbb{T}^m)$ such that for all $f \in \mathcal{U}$, ϕ_f is ergodic and mixing. In case, Λ is connected, openness holds in the C^0-topology.*

In Field & Parry [17], results were obtained on extensions by general compact connected Lie groups over basic sets. For simplicity of exposition we restrict attention to extensions over SRB attractors and make use of results in [15, 16]. We recall that a compact connected Lie group Γ is *semisimple* if and only Γ has finite center.

---

[4]The results of [8] hold for a large class of Γ-extensions over general Anosov systems.

Theorem 3.2. *Let Γ be a compact connected Lie group. Suppose that $\phi : \Lambda \to \Lambda$ is a hyperbolic attractor. Then ϕ_f will be stably SRB for f in a C^0-open and C^∞-dense subset of $C^\infty(\Lambda, \Gamma)$. The same result holds for principal bundle extensions of ϕ.*

   If Γ is semisimple, then ϕ_f is ergodic if and only if it is stably ergodic.

Remark 3.3. Generic stably SRB holds within the class of Γ-equivariant diffeo-morphisms [15]. $\Diamond$

   There are examples [17] where there is generic stable ergodicity of skew ex-tensions over a non-uniformly hyperbolic base.

   The proof of Theorem 3.2 is relatively straightforward when Γ is semisimple. In this case one can exploit the fact that the set of topologically generating pairs for Γ form a (Zariski) open subset of Γ^2 (see [17, 13]).

4. Partially Hyperbolic Symmetric Attractors

Thus far in our examples of stably SRB attractors, we have always assumed that the action of Γ was *free*. We now remove this restriction.

   Throughout this section, we assume that $\Phi : M \to M$ is a smooth Γ-equivariant diffeomorphism of the compact riemannian Γ-manifold M. We also assume that Λ is a compact connected Γ- and Φ-invariant subset of M. We denote the restriction of Φ to Λ by ϕ and let $\tilde{\phi}$ denote the map induced on the orbit space Λ/Γ by ϕ.

   In the context of maps equivariant with respect to a compact connected Lie group, it is useful to make a small change in our definition of attractor. Rather than demanding that $\phi : \Lambda \to \Lambda$ is transitive, we only require that the orbit space map $\tilde{\phi}$ is transitive. With this proviso, our definition of SRB attractor is unchanged.

4.1. Transverse hyperbolicity

When Γ is not finite, we replace our hypothesis of hyperbolicity by one of partial hyperbolicity or *transverse hyperbolicity* [15]. Roughly speaking, we assume that on Λ, ϕ is hyperbolic transverse to Γ-orbits (partial hyperbolicity with center foliation given by Γ-orbits). For this to make sense, we need to assume that all Γ-orbits are of the same dimension. As a further simplification we assume that (a) all Γ-orbits of points in Λ have dimension equal to that of Γ, and (b) that there is an open and dense subset of points of Λ on which Λ acts freely.

Definition 4.1. *The attractor Λ is* transversally hyperbolic *for Φ if*

   (a) *All Γ-orbits in Λ have dimension equal to the dimension of Γ.*
   (b) *There exists a $T\Phi$-invariant splitting $\mathbf{E}^s \oplus \mathbf{E}^u \oplus \mathbf{T}_\Lambda$ of $T_\Lambda M$ into continuous subbundles, and constants $c, C > 0$, $\lambda \in (0,1)$, such that for all $n \in \mathbb{N}$,*

$$\|T_x\Phi^n(v)\| \leq c\lambda^n\|v\|, \quad (v \in \mathbf{E}^s_x, x \in \Lambda) \tag{4.1}$$
$$\|T_x\Phi^n(v)\| \geq C\lambda^{-n}\|v\|, \quad (v \in \mathbf{E}^u_x, x \in \Lambda) \tag{4.2}$$

Example 4.2. Let $f : N \to N$ be a smooth diffeomorphism, and $X \subset N$ be a hyperbolic attractor of f. Every skew extension of f by a compact Lie group is automatically transversally hyperbolic.   ♥

Example 4.3 (Twisted cat map extension). Let $f : \mathbb{T}^2 \to SO(2)$ be a smooth cocycle and consider the corresponding $SO(2)$-extension of the cat map defined in § 3.1:

$$\hat{A}_f : SO(2) \times \mathbb{T}^2 \to SO(2) \times \mathbb{T}^2; (\theta, t) \mapsto (\theta + f(t), \phi(t))$$

Observe that minus the identity map on $\mathbb{R}^2$ induces an involution κ of $\mathbb{T}^2$ which has *four* fixed points. Set $\mathbb{Z}_2(\kappa) = \mathbb{Z}_2$ and note that the cat map $\hat{A}$ is $\mathbb{Z}_2$-equivariant.

We have a *free* action of $\mathbb{Z}_2$ on $SO(2) \times \mathbb{T}^2$ defined by

$$\kappa(\theta, t) = (\theta + \pi, \kappa t)$$

Let $SO(2) \times_{\mathbb{Z}_2} \mathbb{T}^2$ denote the orbit space of the $\mathbb{Z}_2$-action – the *twisted product* of $SO(2)$ and $\mathbb{T}^2$ (by $\mathbb{Z}_2$).

The action of $SO(2)$ on $SO(2) \times \mathbb{T}^2$ drops down to a non-free action on $SO(2) \times_{\mathbb{Z}_2} \mathbb{T}^2$ – there are four singular $SO(2)$-orbits corresponding to the four fixed points of the $\mathbb{Z}_2$-action on $\mathbb{T}^2$.

Suppose that the cocycle $f : \mathbb{T}^2 \to SO(2)$ is $\mathbb{Z}_2$-invariant:

$$f(\kappa t) = f(t), \quad (t \in \mathbb{T}^2).$$

Then $\hat{A}_f$ induces a smooth $SO(2)$-equivariant diffeomorphism $A_f^\star$ of $SO(2) \times_{\mathbb{Z}_2} \mathbb{T}^2$. The volume measure on $SO(2) \times \mathbb{T}^2$ drops down to a volume measure on the twisted product and $A_f^\star$ is volume preserving.

The map $A_f^\star : SO(2) \times_{\mathbb{Z}_2} \mathbb{T}^2 \to SO(2) \times_{\mathbb{Z}_2} \mathbb{T}^2$ provides an example of a partially hyperbolic attractor which is not a skew or principal extension. For this example, we may follow Brin's original quadrilateral argument and perturb within the class of $\mathbb{Z}_2$-invariant cocycles so as to show that $A_f^\star$ is generically stably SRB.   ♥

4.2. Results

We define a map $f : M \to \Gamma$ to be Γ-equivariant if

$$f(\gamma x) = \gamma f(x) \gamma^{-1}, \quad (x \in M, \gamma \in \Gamma).$$

We let $C_\Gamma^\infty(M, \Gamma)$ denote the space of smooth Γ-equivariant maps from M to Γ. Observe that if $f \in C_\Gamma^\infty(M, \Gamma)$, then $\Phi_f = f \circ \Phi : M \to M$ is a Γ-equivariant diffeomorphism. Clearly, Φ_f and Φ induce the same map on the orbit space M/Γ.

The next result was obtained jointly with Matthew Nicol.

Theorem 4.4 ([15]). *Let Γ be a compact connected Lie group, and Λ be a transversally hyperbolic attractor of the smooth Γ-equivariant diffeomorphism of M. For a C^0-open, C^∞-dense subset of maps $f \in C_\Gamma^\infty(M, \Gamma)$, $\phi_f : \Lambda \to \Lambda$ is stably SRB (within the class of equivariant diffeomorphisms on M).*

Remark 4.5. The theorem holds for a wide range of locally maximal transversally hyperbolic sets and for general equilibrium states defined by a Γ-invariant Hölder continuous potential on Λ. We refer to [15] for detailed statements.   ◊

4.3. Notes on the proof of Theorem 4.4.

In general, matters are more complicated than might be expected from the situation described in example 4.3.

Roughly speaking, the proof of Theorem 4.4 depends on showing that Λ/Γ admits 'Markov partitions' and so symbolic dynamics. Using methods based on earlier results of Ruelle & Sullivan [26], we can then prove absolute continuity results for the stable and unstable foliations of Λ and then deduce Livšic regularity theorems that allow the proof of strong results on ergodic components (cf [23, 24]). Finally we are able to apply a theorem of Ledrappier & Young [21] to deduce that Λ is SRB.

As far as the construction of Markov partitions on the orbit space goes, we remark that if the action of Γ is not free, then $\tilde{\phi} : \Lambda/\Gamma \to \Lambda/\Gamma$ is never expansive. In spite of this, it is possible to construct a reasonable (finite) symbolic dynamics on Λ/Γ [15]. The symbolic dynamics turn out to play a crucial role in the construction of equilibrium states – including SRB measures – on Λ [15].

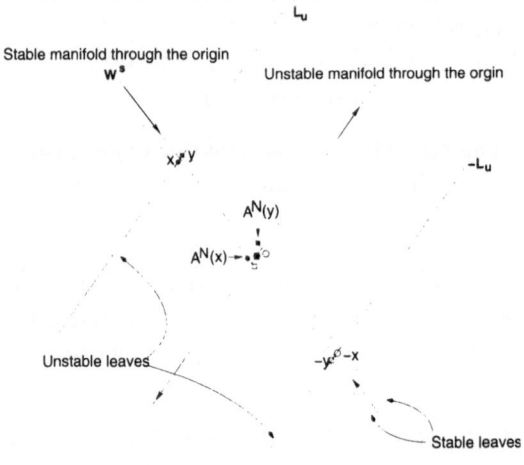

FIGURE 5. Failure of expansiveness

Example 4.6 (cf [11, 15]). The orbit space of the twisted product discussed in example 4.3 is naturally identified with $\mathbb{T}^2/\mathbb{Z}_2$ which is homeomorphic to the 2-sphere S^2. Let $\tilde{A}$ denote the map induced by $A_f^\star$ on S^2[5]. We claim that the induced map $\tilde{A}$ on S^2 is not expansive. In Figure 5, we show the stable and unstable manifolds of the cat map near the fixed point of the $\mathbb{Z}_2$-action corresponding to the origin of $\mathbb{R}^2$. Referring to Figure 5, choose two points x, y which lie on the same unstable leaf L_u and are on opposite sides of, and equidistant along L_u, from

---

[5] $\tilde{\phi}$ is an example of a *pseudo Anosov* map.

the stable manifold W^s through the origin. Let $-x, -y$ denote the $\mathbb{Z}_2$-symmetric images of x, y. It follows by equivariance that $-x, -y \in -L_u$, and by linearity that $x, -y$ lie on the same stable leaf. Let $\varepsilon > 0$. Fixing L_u, we may choose x, y so close to W^s, that for some $N > 0$, we have

$$|A^n(x) - A^n(y)| < \varepsilon, \quad 0 \leq n < N, \quad \text{and } |A^N(x) - A^N(-y)| < \varepsilon.$$

Since A expands L_u and contracts L_s, it follows that for all $n \in \mathbb{Z}$, the distance from $A^n(x)$ to $\{A^n(y), A^n(-y)\}$ is less than ε. Since x, y define different points of the orbit space, it follows that $\tilde{A}$ cannot be expansive. $\heartsuit$

Remark 4.7. The argument given in example 4.6 is quite general and can be shown to apply to any transversally hyperbolic locally maximal set which contains singular group orbits [15]. $\diamondsuit$

References

[1] R Adler, B Kitchens and M Shub. 'Stably ergodic skew products', *Discrete and Continuous Dynamical Systems*, **2** (1996), 349–350 (Errata in **5** (1999), 456).

[2] D V Anosov. 'Geodesic flows on closed Riemann manifolds with negative curvature', *Proc. of the Steklov Inst. of Math.*, **90** (1967).

[3] P. Ashwin and I. Melbourne. Symmetry groups of attractors. *Arch. Rat. Mech. Anal.* **126** (1994) 59–78.

[4] R Bowen. *Equilibrium States and the Ergodic Theory of Anosov Diffeomorphisms*. Springer Lect. Notes in Math. **470**, 1975.

[5] M I Brin. 'Topology of group extensions of Anosov systems', *Mathematical Notes of the Acad. of Sciences of the USSR*, **18**(3) (1975), 858–864.

[6] M I Brin. 'Topological transitivity of one class of dynamic systems and flows of frames on manifolds of negative curvature', *Funkts. Anal. Prilozh*, **9**, No. 1, (1975), 9-19.

[7] M I Brin and J B Pesin. 'Partially hyperbolic dynamical systems', *Math. USSR Izvestia*, **8**, (1974), 177-218.

[8] K Burns and A Wilkinson. 'Stable ergodicity of skew products' *Ann. Sci. de l'Ecole Norm. Sup.* **32** (1999), 859–889.

[9] K Burns, C Pugh, M Shub and A Wilkinson. 'Recent results about stable ergodicity', to appear in *Proc. AMS Summer Research Institute, Seattle, 1999*.

[10] P Chossat and M Golubitsky, *Symmetry increasing bifurcations of chaotic attractors*, Physica D **32** (1988), 423–436.

[11] U-R Fiebig. 'Periodic points and finite group actions on shifts of finite types', *Erg. Thy. & Dynam. Sys.* **13** (1993), 485–514.

[12] M J Field. 'Isotopy and stability of equivariant diffeomorphisms', *Proc. London Math. Soc.* (3) **46** (1983), 487–516.

[13] M J Field. 'Generating sets for compact semisimple Lie groups', *Proc. Amer. Math. Soc.*, (1999), 3361–3365.

[14] M J Field, I Melbourne and M Nicol. 'Symmetric Attractors for Diffeomorphisms and Flows', *Proc. London Math. Soc.* **72**(3) (1996), 657–696.

[15] M J Field and M Nicol. 'Ergodic theory of equivariant diffeomorphisms; Markov partitions and Stable Ergodicity', preprint.

[16] M J Field and V Niţică. 'Stable topological transitivity of skew and principal extensions', *Nonlinearity*, **14** (2001), 1–15.

[17] M J Field and W Parry. 'Stable ergodicity of skew extensions by compact Lie groups', *Topology*, **38**, (1999), 167-187.

[18] M Grayson, C Pugh and M Shub. 'Stably ergodic diffeomorphisms', *Annal of Math.* **140** (1994), 295–329.

[19] A Katok and B Hasselblatt. *Introduction to the Modern Theory of Dynamical Systems* (Encyclopedia of Mathematics and its Applications, **54**, Cambridge University Press, 1995.)

[20] A Katok and A Kononenko. 'Cocycle stability for partially hyperbolic systems', *Math. Res. Lett.* **3** (1996), 191–210.

[21] F Ledrappier and L S Young. 'The metric entropy of diffeomorphisms Part I: Characterization of measures satisfying Pesin's entropy formula', *Ann. of Math* **122** (1985), 509–539.

[22] A N Livšic. 'Cohomology of Dynamical Systems', *Mathematics of the USSR Izvestija*, **6**(6), (1972), 1278-1301.

[23] W Parry. 'Skew-products of shifts with a compact Lie group', *J. London Math. Soc.*, **56**(2) (1997), 400–404.

[24] W Parry and M Pollicott. 'Stability of mixing for toral extensions of hyperbolic systems', *Tr. Mat. Inst. Steklova* **216** (1997), Din Sist. i Smezhnye Vopr., 354–363.

[25] C Pugh and M Shub. 'Stably ergodic dynamical systems and partial hyperbolicity', *J. of Complexity* **13** (1997), 125–179.

[26] D Ruelle and D Sullivan. 'Currents, flows and diffeomorphisms', *Topology*, **14**, (1975), 319-327

[27] M Shub and A Wilkinson. 'Pathological foliations and removable zero exponents', *Invent. Math.*. **139** (2000), 495–508.

[28] L S Young. 'Ergodic theory of chaotic dynamical systems', Lect. Notes in Math. (From *Topology to Computation: Proceedings of the Smalefest*, (eds Hirsch, Marsden, and Shub), Springer-Verlag, New York, Heidelberg, Berlin, 1993.)

[29] P Walters. *An Introduction to Ergodic Theory.* (Springer Verlag, 1982).

[30] A Wilkinson. 'Stable ergodicity of the time-one map of a geodesic flow', *Erg. Th. Dyn. Syst.* **18**(6) (1998), 1545–1588.

[31] R F Williams. One-dimensional non-wandering sets, *Topology* **6** (1967), 473–487.

Michael Field
Department of Mathematics,
University of Houston,
Houston,
Texas, 77204-3476, USA
e-mail: mf@uh.edu

Trends in Mathematics:
Bifurcations, Symmetry and Patterns, 87–99
© 2003 Birkhäuser Verlag Basel/Switzerland

Bistability of Vortex Modes in Annular Thermoconvection

Dan D. Rusu and William F. Langford

Abstract. Spatio-temporal vortex patterns arise in radially forced thermo-convection of a fluid in an annulus. A model based on the two-dimensional Boussinesq fluid equations is analyzed using $O(2)$-equivariant bifurcation theory, coupled with asymptotic numerical analysis. Calculation of the center manifold equations and normal form coefficients allows us to identify and classify the steady-state patterns as well as rotating waves. In particular, we show that the type of secondary transition between the pure mode solutions varies according to the Prandtl number of the fluid.

1. Introduction

The general theory of steady-state mode interactions (of codimension two) in the presence of $O(2)$ symmetry was presented first by Dangelmayr [8] and was discussed further in [2, 10]. Based on this general theory, an analysis of the model-independent aspects of pattern formation in annular convection of a fluid was presented in [15]. This model-independent analysis gives a variety of spatio-temporal patterns near the codimension-two point; for example, a prediction that one of two mutually exclusive transitions would occur between the two primary branches of pure mode solutions in a physically-relevant annular convection model: either a discontinuous transition involving bistability and hysteresis, or a continuous transition along a stable mixed mode branch. Another possibility is the existence of rotating waves, determined by higher order coefficients. It was proposed in [15] that a more detailed description of the model, taking into account physical properties of the fluid and involving explicit numerical computations, would answer the question of which transitions actually occur in a given physical context. That question is resolved in this paper, for annular thermoconvection of a fluid in the Boussinesq approximation, at various Prandtl numbers. We find that the types of transition which occur depend on the Prandtl number: lower Prandtl number fluids such as mercury or gas-giant planetary atmospheres behave differently from higher Prandtl number fluids such as water or air.

The algebraic and analytical methods in [8, 15] have been augmented with numerical algorithms to compute the center manifold and the values of the coefficients of the normal form which decide this dichotomy. Generically, only a small number of modes associated with the PDEs of fluid mechanics go neutrally stable at one time, this number being the codimension of the corresponding point in parameter space. In this paper we consider only codimension-two points involving steady-state modes (corresponding to real zero eigenvalues). The neutrally stable modes define a finite dimensional center manifold at the codimension-two points and allow questions about the behavior of solutions of the original PDE system to be reduced to (local) questions about solutions of an ODE system on this center manifold. An essential feature of this reduction is that the symmetry properties of the solutions are preserved, and as a result the equations on the center manifold take an especially simple form, the equivariant normal form. This normal form determines the bifurcations of various nontrivial solutions, which represent both steady-state and spatio-temporal patterns in the fluid.

The organization of this paper is as follows. Section 2 describes the geophysical systems which motivated this study, together with their governing PDEs and symmetry properties. The neutral stability curves and the codimension-two bifurcation points are determined numerically in Section 3. The primary bifurcations from the trivial solution to nontrivial patterns occur along these neutral stability curves. An asymptotic center manifold reduction is described in Section 4, and implemented numerically, making possible the determination of the coefficients of the normal form. In Section 5, the analytical/numerical analysis of pattern formation in annular thermoconvection is summarized and some of the numerical predictions are presented in the form of bifurcation diagrams.

2. Geophysical Annular Thermoconvection

Geophysical fluid dynamics deals with the motions of the Earth's atmosphere and oceans. This study is motivated in part by geophysical convection occurring in the equatorial plane. Also relevant is equatorial convection in the mantle of planets such as the Earth, and in the thick atmospheres of Jovian planets.

2.1. Thermoconvection in the Equatorial Plane

Consider the Earth as a sphere and the Earth's atmosphere as a spherical shell. Assume the equatorial plane is a plane of reflectional symmetry of the Earth. The equatorial plane intersects the atmosphere in an annulus. The forces acting on the atmosphere normal to this plane are negligible; only tangential forces in the plane are significant. Assuming this idealized symmetry, there is the possibility of flow patterns for which the equatorial plane is invariant. This provides a justification for assuming a three-dimensional fluid to behave as if it were two-dimensional.

The flow in the annulus is convective, driven by the radial gravitational force and buoyancy due to heating at the Earth's surface. This convective force is modelled using the Boussinesq approximation, described in the next subsection. The Coriolis and the centrifugal forces due to the rotation of the planet may be considered negligible in this model [4, 18]. As a consequence of this assumption, the system has $\mathbf{O(2)}$-symmetry and the primary bifurcations are to steady vortex solutions. Recent observational work has revealed unexplained large-scale waves in the equatorial plane of the troposphere [11].

A similar situation occurs in the earth's mantle, which acts as a fluid with an extremely high Prandtl number. The driving force is convective, due to the intense heat at the earth's core. However, the validity of the Boussinesq approximation is questionable for the earth's mantle [20].

2.2. Boussinesq Approximation

Consider a two-dimensional incompressible viscous fluid flow in an annular region Ω, described by the velocity vector field $\mathbf{v}(t, \mathbf{x}) \equiv (u, v)$, where $\mathbf{x} := (r, \theta)$ is position in polar coordinates, at time t. The boundaries of Ω are $r = R_k$ $(k = 1, 2)$, $0 < R_1 < R_2$, or in nondimensional form (with $d := R_2 - R_1$ as the unit of length) $r = r_k$ $(k = 1, 2)$, with $r_1 = \eta/(1-\eta)$ and $r_2 = 1/(1-\eta)$. Here $\eta := R_1/R_2 \in (0, 1)$ is the *radius ratio*. Let $\mathbf{X} := (u, v, \mathrm{T}, p)$ be a vector-valued function of $(t, r, \theta) \in \mathbb{R}_+ \times (r_1, r_2) \times \mathbb{R}/2\pi\mathbb{Z}$, where p is pressure and T is temperature.

Annular thermoconvection may be modeled in a first approximation, and after appropriate scaling, by the nondimensional Boussinesq system [15]

$$\frac{1}{\mathrm{Pr}}\frac{d\mathbf{v}}{dt} = \mathrm{Ra}\left[-\operatorname{grad} p + \nu\Delta\mathbf{v} + (\mathrm{T} - \mathrm{T}_0)\,\mathbf{e}_r\right], \qquad \operatorname{div}\mathbf{v} = 0, \qquad \frac{d\mathrm{T}}{dt} = \Delta\mathrm{T}, \quad (1)$$

considered together with nonslip BCs for the fluid velocity $\mathbf{v}$, and with the temperature T kept fixed at the two boundaries: $\delta\mathrm{T} := \mathrm{T}|_{r=r_1} - \mathrm{T}|_{r=r_2} > 0$. In the above, T_0 denotes a reference value of the temperature, Pr is the *Prandtl number* (an intrinsic physical property of the fluid), and Ra is the *Rayleigh number* (proportional to $\delta\mathrm{T}$). The total time derivative operator $d/dt := \partial/\partial t + \mathbf{v} \cdot \operatorname{grad}$ in the equations is the source of the nonlinearity that makes our problem nontrivial.

There exists an equilibrium satisfying $(\mathbf{v}, \partial/\partial t) \equiv \mathbf{0}$, with the heat transported purely by thermal conduction. This motionless radially conducting solution $\mathbf{X}_e = (0, 0, p_e(r), \mathrm{T}_e(r))$ of the Boussinesq system may be determined analytically: the temperature at equilibrium is given by $\mathrm{T}_e(r) = (\delta\mathrm{T}_0/\ln\eta)\ln(r/r_1) + \mathrm{T}_0$.

2.3. Symmetries

The annular geometry and the Euclidean symmetry of the Boussinesq system (1) imply that our system is invariant under the action of the continuous group $\mathbf{O(2)}$, generated by the special orthogonal subgroup $\mathbf{SO(2)}$ consisting of planar rotations R_φ, $\varphi \in [-\pi, \pi]$, and the finite subgroup $\mathbf{Z_2} = \mathbf{Z_2}(\kappa)$, where κ is the reflection about the x-axis.

A finite subgroup of $\mathbf{O(2)}$ of interest is the dihedral group $\mathbf{D}_n$, identified with the group of 2×2 matrices generated by $R_{2\pi/n}$ $(n \in \mathbb{N})$ and κ. In the asymptotic limit of a small gap the geometry becomes that of flow between two parallel walls, with the symmetry $\mathbf{Z_2} \oplus \mathbf{Z_2}$ and periodic BCs.

3. Linear Stability Analysis

The first step of the analysis is to determine in the (η, Ra) plane the *neutral stability curves* which separate the domains where the equilibrium solution is linearly stable from those where it is linearly unstable. Off of these neutral curves, hyperbolicity implies that the nonlinear system and its linearization have locally the same asymptotic stability/instability, for the equilibrium solution.

3.1. Nonlinear Initial-Boundary Value Problem

Translate the equilibrium solution of the Boussinesq system to the origin (trivial solution). Then the system may be written in the form of a nonlinear initial-boundary value problem, with homogeneous BCs, of the form

$$
\begin{aligned}
\partial_t \mathbf{CX} &= \mathbf{L}(\mu)\mathbf{X} + \mathbf{N}(\mathbf{X}, \mathbf{X}; \mu) \quad \text{in} \quad \Omega, \\
\mathbf{CX}|_{t=0} &= \mathbf{X}_0 \quad \text{in} \quad \overline{\Omega}, \\
\mathbf{CX} &= \mathbf{0} \quad \text{on} \quad \partial\Omega.
\end{aligned}
\tag{2}
$$

Here $\mu := (\mathrm{Ra}, \eta)$, $\mathbf{C} := \mathrm{diag}(1, 1, 1, 0)$, $\mathbf{L}(\mu)$ is a linear differential operator with non-constant coefficients, and $\mathbf{N}(\,.\,,\,.\,; \mu)$ is a quadratic operator satisfying $\mathbf{N}(\mathbf{0}, \mathbf{0}; \mu) = \mathbf{0}$. It includes the advection terms derived from the material time derivative of $\mathbf{CX}$, see [20].

The pressure variable p may be eliminated from the system by any of several techniques, see [20]. Here, we consider pressure as a Lagrange multiplier associated with a constraint of the form of the divergence-free condition, and thereby maintain p in the system. Numerically, this leads to treating p as a secondary variable defined on a mid-point grid, where also the incompressibility constraint is discretised [17, 19]. The use of a mid-point grid stabilizes the numerical method; if pressure is calculated at the same points as velocities, then there is a risk of unstable numerical solutions with an "oscillatory pressure". In addition, it avoids the need for BCs for the pressure.

3.2. Normal Modes

The PDE system (2) has a linearization of the form $\partial_t \mathbf{CX} = \mathbf{L}(\mu)\mathbf{X}$. The substitution $\mathbf{X}(t, r, \theta) = e^{\lambda t}\widehat{\mathbf{X}}(r, \theta)$ reduces this to a generalized eigenvalue problem $\mathbf{L}(\mu)\widehat{\mathbf{X}} = \lambda\,\mathbf{C}\widehat{\mathbf{X}}$ with homogeneous BCs. The general solution is then a superposition of *normal modes* of the form $\widehat{\mathbf{X}}_m(r, \theta) = \mathbf{Y}_m(r) \otimes \Xi(m\theta)$, where $\otimes$ indicates dyadic product, $\mathbf{Y}_m := (U, V, T, P)$, $\Xi(\alpha) := \left(e^{i\alpha}, -ie^{i\alpha}, e^{i\alpha}, e^{i\alpha}\right)$, and U, V, T, P are real-valued functions. The linear stability computations (which are Prandtl independent) are thus reduced to solving a linear two-point boundary value problem for $\mathbf{Y}_m(r)$, $r_1 \leq r \leq r_2$.

3.3. Neutral Stability Curves

The generalized eigenvalue problem with homogeneous BCs yields a countable infinity of neutral curves $\mathrm{Ra} = \mathrm{Ra}^{(m)}(\eta)$, $m \in \mathbb{N}$, where an eigenvalue is zero. The *bicritical point* (η_c, Ra_c) is the transverse intersection of two consecutive neutral curves, $\mathrm{Ra} = \mathrm{Ra}^{(m)}(\eta)$ and $\mathrm{Ra} = \mathrm{Ra}^{(l)}(\eta)$, $l = m+1$. These were computed using finite differences on a staggered mesh, for mode numbers $m = 2, \ldots, 15$. The mesh size was successively refined to test for numerical convergence. All computations were implemented in Matlab. Neutral stability curves corresponding to consecutive values of the mode number m are shown in Figure 1. For example, with $(m,l) = (2,3)$ the codimension-two point is $(\eta_c, \mathrm{Ra}_c) = (0.2633, 1896)$ [20]. All of the results in Figure 1 are independent of Prandtl number. Note that the neutral stability curves become successively closer as m gets bigger.

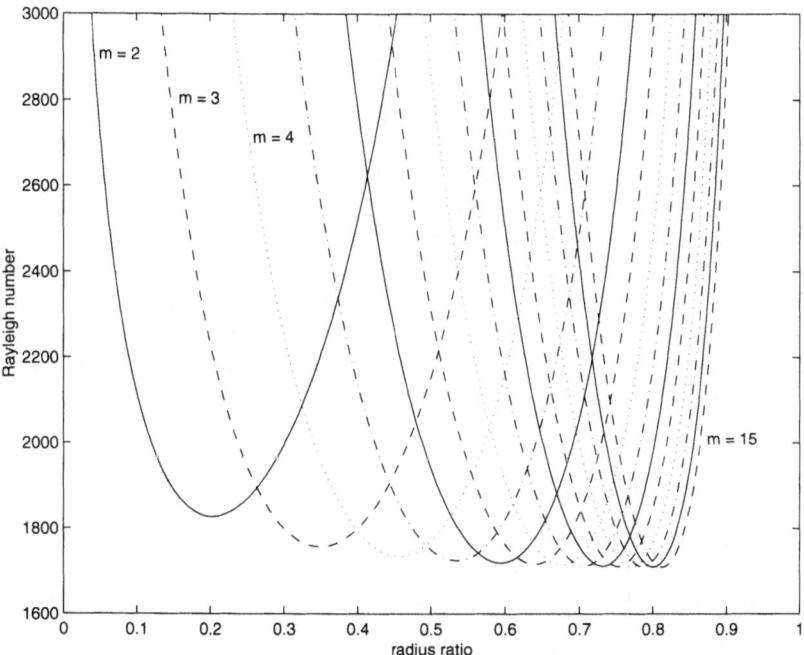

FIGURE 1. Consecutive Neutral Stability Curves. The trivial solution is stable at all points below the envelope of these curves.

4. Center Manifold Reduction

Under reasonable conditions satisfied by many PDE systems such as this one [12], there exists a *center manifold* W^c, tangent to the center eigenspace at the origin, and of the same dimension (four-dimensional at these codimension-two points). If the remainder of the spectrum is bounded away from the imaginary axis in the left half-plane, as in the cases considered here, then locally all solutions near W^c will approach it exponentially with increasing time. Thus the study of the long-time behavior of solutions of the PDE system near the origin is reduced to that of solutions of an ODE system on this center manifold. Symmetry properties are preserved by this reduction.

We use an asymptotic method for extracting amplitude equations for the ODE system on the center manifold from the PDE model of the physical system, when the system is near points of codimension-two bifurcation. The solutions on the center manifold are computed approximately using Taylor series about the bicritical point. See [20] for a detailed description of this reduction procedure. It requires solving a sequence of linear nonhomogeneous two-point boundary value problems, similar to the generalized eigenvalue problems in the previous section. Similar ideas have been used in [3, 6, 19], for other bifurcation problems.

4.1. Numerical Analysis

In order to solve the sequence of linear nonhomogeneous boundary value problems at each step of the asymptotic procedure, we consider again a *staggered mesh*. All computations were performed twice, employing independently both a spectral method and a finite difference method, as a test of accuracy. The results obtained in the two approaches agreed very well [20].

Our final conclusions, based on equivariant bifurcation theory, require only the knowledge of the signs of several quantities defined in terms of the numerically computed coefficients of the normal form. Small errors in the computations will not affect the signs of these coefficients; this implies that our qualitative predictions are robust to the inherent numerical errors.

5. O(2) Mode Interactions

The general context for a symmetry based analysis of pattern formation is *equivariant bifurcation theory*, presented for the first time as a complete theory in [10]. The application of this theory to annular convection was summarized in [15].

5.1. O(2) Equivariant Normal Form

The action of $\mathbf{O(2)}$ on $\mathbb{C}^2$ is given by: $\theta \cdot (z_1, z_2) = (e^{im\theta} z_1, e^{il\theta} z_2)$, $\kappa \cdot (z_1, z_2) = (\bar{z}_1, \bar{z}_2)$, where $\theta \in [0, 2\pi)$. The normal form for the $\mathbf{O(2)}$-equivariant ODE system is

$$\begin{bmatrix} \dot{z}_1 \\ \dot{z}_2 \end{bmatrix} = p_1 \begin{bmatrix} z_1 \\ 0 \end{bmatrix} + p_2 \begin{bmatrix} 0 \\ z_2 \end{bmatrix} + q_1 \begin{bmatrix} \bar{z}_1^{l-1} z_2^m \\ 0 \end{bmatrix} + q_2 \begin{bmatrix} 0 \\ z_1^l \bar{z}_2^{m-1} \end{bmatrix} \quad (3)$$

together with the two complex conjugate equations for $\bar{z}_1$ and $\bar{z}_2$. Here p_1, p_2, q_1, q_2 are invariant real functions of $(u, v, w; \mu)$, such that $p_1(0,0,0) = p_2(0,0,0) = 0$, and $\{u, v, w\}$ is a Hilbert basis, with $u := |z_1|^2$, $v := |z_2|^2$, $w := z_1^l \bar{z}_2^m + \bar{z}_1^l z_2^m$.

The *lattice of isotropy subgroups* $(m \geqslant 2)$ provides a valuable way to organize conceptually the bifurcation analysis. For our case, the initial bifurcations lead to waves with maximal isotropy subgroups $\mathbf{D}_m$ or $\mathbf{D}_l$; the secondary instabilities of these waves lead further to motions with submaximal subgroup $\mathbf{Z_2}$. Each isotropy subgroup determines a corresponding fixed-point subspace, which is invariant for the time-evolution of solutions. Since $\mathrm{Fix}(\mathbf{Z_2})$ contains both $\mathrm{Fix}(\mathbf{D}_m)$ and $\mathrm{Fix}(\mathbf{D}_l)$

and there are no steady-states of (3) other than in Fix($\mathbf{Z_2}$), we may restrict the system to Fix($\mathbf{Z_2}$) and solve it there. Thus symmetry has further reduced the dimension of the problem for steady-states from four to two. However, since interesting dynamics (such as rotating waves) may be expected to occur off of the subspace Fix($\mathbf{Z_2}$) $\simeq \mathbb{R}^2$ of $\mathbb{C}^2$, amplitude-phase equations also are employed.

Based on the experimental evidence, and confirmed by our numerical calculations, the codimension-two points of neutral stability involve only *consecutive* mode numbers as shown in Figure 1. Therefore, we take $l = m + 1$ with $m \geqslant 2$ in the following.

5.2. $\mathbf{Z_2} \oplus \mathbf{Z_2}$ Symmetric Approximation

For large m, the m-dependent terms are higher order, and the resulting equations are essentially equivalent to the $\mathbf{Z_2} \oplus \mathbf{Z_2}$-symmetric normal form [10, 15]. There are two primary pitchfork branches from the trivial solution, and secondary mixed mode branches with both of $x, y \neq 0$, occurring as four conjugates ($\pm x, \pm y$) under the $\mathbf{Z_2} \oplus \mathbf{Z_2}$ symmetry. The secondary $\mathbf{Z_2}$-branches may link the two primary branches in two different ways. Either the secondary branch is itself unstable but effects an exchange of stabilities with *bistability* and the possibility of *hysteresis* between the two primary branches, or the secondary branch is itself stable and effects a *continuous transition* between the primary branches, along a stable mixed mode branch. Which of these two mutually exclusive possibilities actually occurs is determined by the (numerically computed) normal form coefficients. In the present study, the leading order normal form coefficients were computed for (3) in the case $q_1 = q_2 = w = 0$ (the $\mathbf{Z_2} \oplus \mathbf{Z_2}$ symmetric approximation), for m ranging from 2 to 15. In all cases, the type of secondary bifurcation found was that of bistability/hysteresis between primary branches, and not a continuous transition along a stable mixed mode branch, see [20].

The m-dependent terms in the normal form (3) break the $\mathbf{Z_2} \oplus \mathbf{Z_2}$ symmetry, as described in [15]. A single $\mathbf{Z_2}$-equivariance remains: $(x, y) \to (-x, y)$ if m odd, and $(x, y) \to (x, -y)$ if m is even. The effects of this can be seen in the "imperfect" pitchfork bifurcations in the bifurcation diagrams below. Another consequence of this broken symmetry is that the four-dimensional normal form reduces only to a three-dimensional amplitude-phase system for time-dependent solutions.

5.3. Normal Form Restricted to Fix($\mathbf{Z_2}$)

With $l = m + 1$ in (3), the flow on the center manifold is approximated by the truncated system (retaining leading order terms in p_1, p_2, q_1, q_2)

$$\begin{aligned}
\dot{z}_1 &= (\alpha + a_{30}\, u + a_{12}\, v)\, z_1 + a_{22}\, \bar{z}_1^m z_2^m, \\
\dot{z}_2 &= (\beta + b_{21}\, u + b_{03}\, v)\, z_2 + b_{31}\, z_1^{m+1} \bar{z}_2^{m-1},
\end{aligned} \qquad (4)$$

together with the conjugate equations. For the numerical computations, we have taken

$$\begin{aligned}
a_{22}\, b_{31} &\neq 0, \quad \text{for} \quad m = 2 \quad \text{(strong mode interactions)}, \\
a_{22} = b_{31} &= 0, \quad \text{for} \quad m \geqslant 4 \quad \text{(weak mode interactions)}.
\end{aligned}$$

Note that the approximation $a_{22} = b_{31} = 0$ in (4) reduces the system to the third order $\mathbf{Z_2} \oplus \mathbf{Z_2}$-equivariant bifurcation problem on $\mathbb{R}^2$ described above. For large m, these terms cause only small symmetry-breaking perturbations of the $\mathbf{Z_2} \oplus \mathbf{Z_2}$ case. For $m \geqslant 4$, these symmetry-breaking terms are of order at least 8. For $m = 3$, these terms are order 6; just one higher than the symmetric fifth order terms in $p_1 z_1$ and $p_2 z_2$ of (3), which have been omitted from (4). We have not computed the normal form to order 6 in the case $m = 3$; however, we do present here, for the first time, the effects of the $\mathbf{Z_2} \oplus \mathbf{Z_2}$ symmetry-breaking terms in the case $m = 2$.

The truncated bifurcation equations restricted to $\mathrm{Fix}(\mathbf{Z_2})$ are

$$x \left(\alpha + a_{30}\, x^2 + a_{12}\, y^2 + a_{22}\, x^{m-1}\, y^m \right) = 0, \qquad (5)$$
$$y \left(\beta + b_{21}\, x^2 + b_{03}\, y^2 + b_{31}\, x^{m+1}\, y^{m-2} \right) = 0.$$

where we assume $a_{30}\, b_{03} \neq 0$, $\Delta := a_{30}b_{03} - a_{12}b_{21} \neq 0$. The steady-state solutions for the normal form restricted to $\mathrm{Fix}(\mathbf{Z_2})$ are:

• *Trivial solution* T, with $x = y = 0$.
• *Pure mode solutions* S_m and S_l given by primary bifurcations from T: $\mathrm{S}_m : x^2 = -\alpha/a_{30}$, $y = 0$, and $\mathrm{S}_l : x = 0$, $y^2 = -\beta/b_{03}$, respectively. Note that S_m exists if and only if $a_{30}\,\alpha < 0$, and S_l exists if and only if $b_{03}\,\beta < 0$.
• *Mixed mode solutions* $\mathrm{S}_\pm$ arising in pairs as secondary bifurcations from S_m or S_l, with both $x, y \neq 0$, given when $m = 2$ by

$$\alpha_m - \Delta\, y^2 + x \left(-a_{30}\, b_{31} x^2 + a_{22}\, b_{21} y^2 \right) = 0, \qquad (6)$$
$$\beta_l - \Delta\, x^2 + x \left(a_{12}\, b_{31} x^2 - a_{22}\, b_{03} y^2 \right) = 0,$$

where $\alpha_m := b_{21}\, \alpha - a_{30}\, \beta$ and $\beta_l := -b_{03}\, \alpha + a_{12}\, \beta$. Note that $\mathrm{S}_\pm$ exist if and only if $\Delta\, \alpha_m > 0$ and $\Delta\, \beta_l > 0$.

The corresponding *bifurcation points* lie on the following lines in parameter space (see Figures 2 and 3)

$$\mathrm{S}_m \cap \mathrm{T}: \ \alpha = 0, \ \text{and} \ \mathrm{S}_l \cap \mathrm{T}: \ \beta = 0, \ \text{and} \ \mathrm{S}_\pm \cap \mathrm{S}_l: \ \beta_l = 0,$$
$$\mathrm{S}_\pm \cap \mathrm{S}_m: \ \alpha_m = 0, \ \text{if} \ m \geqslant 4 \ \text{and} \ \alpha_m \pm \mathcal{O}(3/2) = 0, \ \text{if} \ m = 2.$$

Transversality implies the persistence of these solutions (in the parameter-phase space) and of the stability assignments, when the higher-order terms are reintroduced in the normal form.

5.4. Amplitude-Phase Equations

In this section we work with general l and m, although our main interest is the case $l = 3$, $m = 2$. With $z_1 = re^{i\varphi}$, $z_2 = se^{i\psi}$, and $\theta := l\varphi - m\psi$ in (3), we obtain the reduced amplitude-phase system

$$\dot{r} = r\, p_1 + r^{l-1} s^m q_1 \cos\theta, \quad \dot{s} = s\, p_2 + r^l s^{m-1} q_2 \cos\theta, \qquad (7)$$
$$\dot{\theta} = -r^{l-2} s^{m-2} \left(l s^2 q_1 + m r^2 q_2 \right) \sin\theta,$$

which is a closed system, since $r^2 q_2\, \dot{\varphi} + s^2 q_1\, \dot{\psi} \equiv 0$ [8, 10]. The emergence of the mixed phase θ and the corresponding reduction is a consequence of the $\mathbf{SO(2)}$-symmetry. Note that $\theta = 0$ and $\theta = \pi$ are invariant planes under the flow.

The following steady-states of the amplitude-phase system with $\theta = 0, \pi$ ($\dot{\varphi} = 0, \dot{\psi} = 0$) correspond to true steady-states of the original normal form: *pure modes* S_m: $s = 0, r \neq 0$, S_l: $r = 0, s \neq 0$, and *mixed modes* $S_\pm$, when $r \neq 0, s \neq 0$ and $\sin \theta = 0$, $p_1 \pm r^{l-2} s^m q_1 = 0$, $p_2 \pm r^l s^{m-2} q_1 = 0$.

Steady-state solutions of the amplitude-phase system with $\theta \neq 0$, $\theta \neq \pi$ are determined by $ls^2 q_1 + mr^2 q_2 = 0$, and $p_1 + r^{l-2} s^m q_1 \cos \theta = 0$, $p_2 + r^l s^{m-2} q_2 \cos \theta = 0$. They represent quasiperiodic solutions of the original normal form (*rotating waves* RW) in which θ remains constant, but φ and ψ both increase/decrease linearly with time. They come in pairs and exist if $q_1 q_2 < 0$, which locally holds in the case $m = 2$, $l = 3$, if

$$a_{22} b_{31} < 0. \tag{8}$$

They are examples of *relative equilibria*: "rigidly rotating" motions which become true equilibria in a suitable rotating coordinate frame. Their isotropy group is the *spatio-temporal subgroup* $\widetilde{SO}(2) \subset SO(2) \times S^1$, where $SO(2) \subset O(2)$ acts on $\mathbb{C}^2$ by an m-fold rotation on the first component, an l-fold rotation on the second component, and the circle group S^1 stems from the temporal oscillations [8, 10]. The stabilities of these steady-state solutions are determined in [20].

5.5. Bifurcation Predictions

From the numerically computations of the coefficients of the normal form, we obtain a complete classification of the types of steady-state solutions (and some time dependent solutions). The case $m = 1$ is quite different from the others [2] and will not be discussed here. We consider only two cases: $m = 2$ (strong mode interactions) and $m \geqslant 4$ (weak mode interactions). The intermediate mode interaction $m = 3$ was not considered because of the complexity of the calculations involved in computing the normal form to sixth order. For brevity, we present here the bifurcation diagrams only for the $(m, l) = (2, 3)$ case.

In our calculations: $\alpha := a_{1\varepsilon} \varepsilon + a_{1\varkappa} \varkappa$ and $\beta := b_{\varepsilon 1} \varepsilon + b_{\varkappa 1} \varkappa$, where the bifurcation parameters are ε (corresponding to the temperature gradient or Rayleigh number) and $\varkappa$ (corresponding to the annular radius ratio). In the $\varkappa, \varepsilon$–plane we have (see Figures 2 and 3):

- *primary bifurcation lines*: i) (PLm): $\alpha = 0$ and ii) (PLl): $\beta = 0$, both independent of the Prandtl number.
- *secondary bifurcation half-lines* (in the existence domain of S_m and S_l): iii) (SLm): $\alpha_m = 0$ and iv) (SLl): $\beta_l = 0$. Actually for $m = 2$, the line (SLm) no longer leads to the simultaneous bifurcation of S_+ and S_-, but is split into iii) (SLm)$_\pm$: $\alpha_m \pm \mathcal{O}(3/2) = 0$. These curves are infinitesimally close to each other and terminate tangential to the half-line (SLm) at the origin.

From our numerical calculations $a_{30} < 0$, $b_{03} < 0$ and $a_{1\varepsilon} > 0$, $b_{\varepsilon 1} > 0$, see [20], we obtain two supercritical pitchfork branches S_m existing for $\alpha > 0$ and S_l existing for $\beta > 0$. Since $\Delta < 0$, $S_\pm$ exist if and only if $\alpha_m < 0$ and $\beta_l < 0$. These mixed modes bifurcate from the pure modes along the secondary bifurcation half-lines (SLm) and (SLl), located in the existence domain of the pure modes.

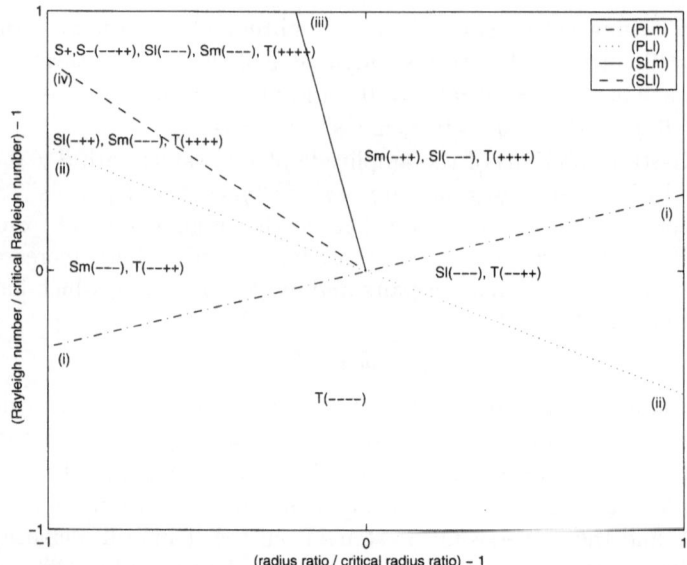

FIGURE 2. Bifurcation Lines and Stability Assignments for $(m, l) = (2, 3)$: Mercury (Pr $= 0.027$).

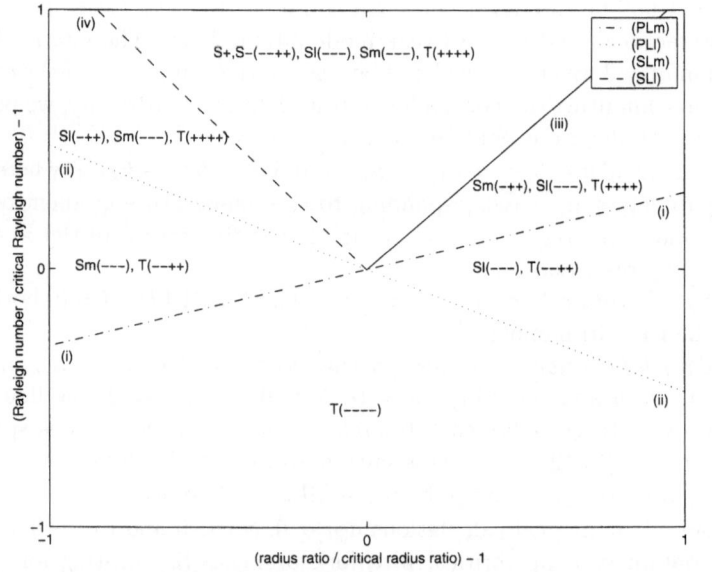

FIGURE 3. Bifurcation Lines and Stability Assignments for $(m, l) = (2, 3)$: Air (Pr $= 0.7$) and Water (Pr $= 7.0$).

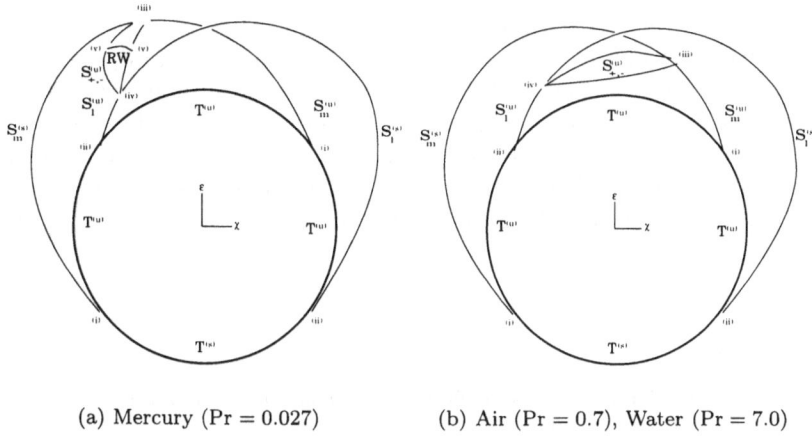

(a) Mercury (Pr = 0.027)   (b) Air (Pr = 0.7), Water (Pr = 7.0)

FIGURE 4. Gyrant Bifurcation Diagrams, for $(m, l) = (2, 3)$.

The stability of $S_\pm$ is determined by the eigenvalues $\{\ 0,\ e_\pm,\ (\lambda_1)_\pm,\ (\lambda_2)_\pm\ \}$. $S_\pm$ are unstable since $(\lambda_1)_\pm$ and $(\lambda_2)_\pm$ are real and of opposite sign. The eigenvalue $e_\pm$ will change sign if and only if condition (8) on fourth order coefficients holds in the existence domain of $S_\pm$, which implies the existence of RW $\cap\ S_\pm$. We find that $S_\pm$ may encounter a bifurcation to a pair of RW only for the case of a very low Prandtl number fluid, such as mercury (see Figures 4(a), 4(b)). We conjecture that RW is unstable. We conjecture further that these RW may appear also in the case of the gas-giant planets (which are believed to be very low Prandtl number fluids) but not in the Earth's atmosphere.

5.6. Bifurcation Diagrams

We collect all pieces of information together in *gyrant bifurcations diagrams* with $(\varepsilon, \varkappa)$ moving in a circular path around a circle (the trivial solution) centered at the bicritical point, see Figures 4(a)(b).

The superscript (s) indicates the stable branches and the superscript (u) indicates the unstable branches. Labelling of bifurcation points is consistent in all the figures and the label key is (i): bifurcation of m-mode from the trivial solution; (ii): bifurcation of l-mode from the trivial solution; (iii): bifurcation of mixed modes from the m-mode; (iv): bifurcation of mixed modes from the l-mode; (v): bifurcation of rotating waves from the mixed modes.

6. Conclusions

We draw four conclusions from the computations on which Figures 2, 3 and 4(a)(b) are based. First, at all Prandtl numbers, the case of bistability/hysteresis between the primary branches occurs, and not the case of a stable mixed mode branch joining the primary branches. Second, the region of bistability exists only for radius ratios to the left of the codimension-two point when the Prandtl number is low, but extends to both sides for higher Prandtl numbers. Third, we predict the existence of rotating waves in the case of lower Prandtl number fluids (such as mercury, and possibly gas-giant planetary atmospheres). Finally, in the case of the earth's atmosphere (high Prandtl number and very large mode number m), our calculations are consistent with the coexistence of stable large scale waves or vortices in the equatorial plane, phase-shifted systematically with longitude, as observed for example in [11].

References

[1] A. Alonso, M. Net and E. Knobloch, *On the transition to columnar convection*, Phys. Fluids A, **7** (1995), 935–940.

[2] D. Armbruster, J. Guckenheimer and P. Holmes, *Heteroclinic cycles and modulated travelling waves in systems with* **O**(2) *symmetry*, Physica D, **29** (1988), 257–282.

[3] A. Arneodo, P.H. Coullet and E.A. Spiegel, *The dynamics of triple convection*, Geophys. Astrophys. Fluid Dynamics, **31** (1985), 1–48.

[4] B.W. Atkinson, *Meso-scale Atmospheric Circulations*, Academic Press, London, 1981.

[5] P. Chossat and G. Iooss, *The Couette–Taylor Problem*, Springer–Verlag, New York, 1994.

[6] P. Coullet and E.A. Spiegel, *Amplitude equations for systems with competing instabilities*, SIAM J. Appl. Math., **43** (1983), 776–821.

[7] J.D. Crawford and E. Knobloch, *Symmetry and symmetry-breaking bifurcations in fluid dynamics*, Annu. Rev. Fluid Mech., **23** (1991), 341–387.

[8] G. Dangelmayr, *Steady-state mode interactions in the presence of* **O**(2) *symmetry*, Dynam. Stab. Syst., **1** (1986), 159–185.

[9] M. Golubitsky and W.F. Langford, *Pattern formation and bistability in flow between counterrotating cylinders*, Physica D, **32** (1988), 362–392.

[10] M. Golubitsky, I. Stewart and D.G. Schaeffer, *Singularities and Groups in Bifurcation Theory, Vol. 2*, Springer–Verlag, New York, 1988.

[11] K. Hamilton, *Observation of an ultraslow large-scale wave near the tropical tropopause*, J. Geophys. Res., **102** (1997), 457–464.

[12] D. Henry, *Geometric Theory of Semilinear Parabolic Equations*, Springer–Verlag, New York, 1981.

[13] E. Knobloch and J. Guckenheimer, *Convective transitions induced by a varying aspect ratio*, Phys. Rev. A, **27** (1983), 408–417.

[14] E. Knobloch, *Bifurcations in rotating systems*, Lectures on Solar and Planetary Dynamics, Edited by M.R.E. Proctor and A.D. Gilbert, Cambridge University Press (Cambridge), (1994), 331–370.

[15] W.F. Langford and D.D. Rusu, *Pattern formation in annular convection*, Physica A, **261** (1998), 188–203.

[16] W.F. Langford, R. Tagg, E.J. Kostelich, H.L. Swinney and M. Golubitsky, *Primary instabilities and bicriticality in flow between counter-rotating cylinders*, Phys. Fluids, **31** (1988), 776–785.

[17] S.V. Patankar, *Numerical Heat Transfer and Fluid Flow*, Hemisphere Publishing Corporation, Washington, 1980.

[18] J. Pedlosky, *Geophysical Fluid Dynamics*, Springer–Verlag, New York, 1987.

[19] A.J. Roberts, *Low-dimensional Modelling in Hydrodynamics*, University of Southern Queensland, Toowoomba, 1995.

[20] D.D. Rusu, *Pattern Formation in Annular Convection: An Equivariant Bifurcation Analysis*, Ph.D. thesis, University of Guelph, Ontario, 2000.

[21] A. Vanderbauwhede and G. Iooss, *Center manifold theory in infinite dimensions*, Dynam. Report., **1** (1992), 125–163.

Dan D. Rusu, William F. Langford
Department of Mathematics and Statistics
University of Guelph
Guelph, Ontario N1G 2W1, Canada
e-mail: `wlangfor@uoguelph.ca`

Trends in Mathematics:
Bifurcations, Symmetry and Patterns, 101–114
© 2003 Birkhäuser Verlag Basel/Switzerland

Secondary Instabilities of Hexagons:
A Bifurcation Analysis of Experimentally
Observed Faraday Wave Patterns

A.M. Rucklidge, M. Silber, and J. Fineberg

Abstract. We examine three experimental observations of Faraday waves generated by two-frequency forcing, in which a primary hexagonal pattern becomes unstable to three different superlattice patterns. We analyse the bifurcations involved in creating the three new patterns using a symmetry-based approach. Each of the three examples reveals a different situation that can arise in the theoretical analysis.

1. Introduction

The classic Faraday wave experiment consists of a horizontal layer of fluid that spontaneously develops a pattern of standing waves on its surface as it is driven by vertical oscillation with amplitude exceeding a critical value. Recent experiments have revealed a wide variety of complex patterns, particularly in the large aspect ratio regime and with a forcing function containing two commensurate frequencies [1, 2, 3]. Transitions from the flat surface to a primary, spatially periodic, pattern can be studied using equivariant bifurcation theory [4]. These group theoretic techniques may also be applied to secondary spatial period-multiplying transitions to patterns with two distinct spatial scales (so called *superlattice* patterns) as demonstrated by Tse *et al.* [5].

We apply the method of Tse *et al.* [5] to the analysis of three superlattice patterns observed when secondary subharmonic instabilities destroy the basic hexagonal standing wave pattern in two-frequency Faraday wave experiments. We can make use not only of the general symmetry-based approach from [5] but also of many of the detailed results. The reason for this is that in their paper, Tse *et al.* considered instabilities of hexagonal patterns that broke the translation symmetry of the hexagons, but that remained periodic in a larger hexagonal domain comprising twelve of the original hexagons. The instabilities under consideration here satisfy exactly the same conditions (though in fact they remain periodic in smaller domains as well).

We begin by specifying the coordinate system and symmetries we will use in section 2, then describe the symmetries of the three experimental patterns in

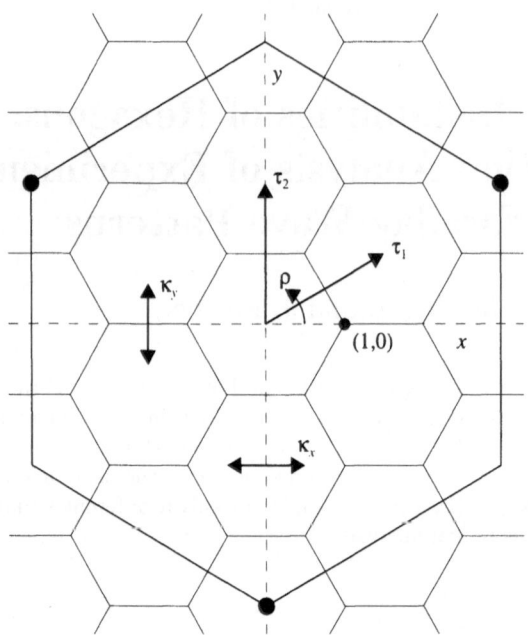

FIGURE 1. The coordinate system and certain elements of the symmetry group Γ. The origin of the coordinate system is at the centre of the
diagram, and the point $(1,0)$ is indicated. The small hexagons represent
the primary pattern, which is invariant under reflections (κ_x and κ_y),
60° rotations (ρ) and translations (τ_1 and τ_2). The secondary patterns
are all periodic in the larger hexagonal box. The three corner points
labelled with solid circles are identified through the assumed periodicity.

section 3. In section 4, we apply Tse *et al.*'s method of analysis to these three
patterns, and present normal forms and stability calculations in section 5. We
conclude in section 6.

2. Coordinates and symmetries

The primary pattern is made up of regular hexagons, which are invariant under
the group D_6 (made up of 60° rotations and reflections) combined with translation
from one hexagon to the next (see figure 1). Tse *et al.* [5] studied experimental
patterns reported in [6], which had the feature that after the secondary instability,
the pattern remained periodic in the larger hexagonal box in figure 1. The 144-
element spatial symmetry group of the primary hexagonal pattern within this box

is Γ, generated by the following reflection κ_x, rotation ρ and translations τ_1 and τ_2:

$$\kappa_x : (x, y) \rightarrow (-x, y) \qquad\qquad \tau_1 : (x, y) \rightarrow (x, y) + \left(\frac{3}{2}, \frac{\sqrt{3}}{2}\right) \quad (1)$$

$$\rho : (x, y) \rightarrow \left(\frac{1}{2}x - \frac{\sqrt{3}}{2}y, \frac{\sqrt{3}}{2}x + \frac{1}{2}y\right) \qquad \tau_2 : (x, y) \rightarrow (x, y) + \left(0, \sqrt{3}\right) \quad (2)$$

We also define $\kappa_y = \kappa_x \rho^3$, and note the following identities:

$$\kappa_x^2 = \kappa_y^2 = \rho^6 = \tau_1^6 = \tau_2^6 = \tau_1^2 \tau_2^2 = \text{identity}, \qquad\qquad (3)$$

$$\rho \kappa_x = \kappa_x \rho^5, \qquad \tau_1 \kappa_x = \kappa_x \tau_1^5 \tau_2, \qquad \tau_2 \kappa_x = \kappa_x \tau_2, \qquad\qquad (4)$$

$$\tau_1 \rho = \rho \tau_1^3 \tau_2, \qquad \tau_2 \rho = \rho \tau_1, \qquad \tau_1 \tau_2 = \tau_2 \tau_1. \qquad\qquad (5)$$

The time translation τ_T advances time by one period T of the forcing function, which is the same as the temporal period of the hexagonal pattern. This time translation is combined with the spatial symmetries above to give spatio-temporal symmetries.

3. Experimental patterns

The three experimentally observed patterns are shown in figure 2(a-c), visualised using the techniques described in [7]. Patterns (a) and (b) are both obtained using Dow-Corning silicone oil with viscosity 47 cSt and layer depth 0.35 cm, while pattern (c) was found using a 23 cSt oil layer of depth 0.155 cm. All three patterns are obtained with forcing function containing two frequencies in the ratio 2 : 3; pattern (a) is found with frequencies 50 and 75 Hz, pattern (b) with frequencies 70 and 105 Hz, and pattern (c) with 40 and 60 Hz driving frequencies. Pattern (c) was reported previously in [7]. Typically, the secondary bifurcations occur at forcing amplitudes between 10 and 50% larger than the critical acceleration for the primary hexagonal state. Further experimental details can be found in [7, 8].

For the purposes of the analysis, we consider the idealised versions of these experimental patterns, shown in figure 2(d-f). The first pattern in figure 2(a,d) retains the D_6 symmetry of the original hexagons but breaks certain translation symmetries. It is periodic in the medium-sized dashed hexagon in figure 2(d), which implies that the pattern is invariant under the translations τ_1^3 and $\tau_1 \tau_2$. It has no spatio-temporal symmetries. The second pattern is similar, although it possesses only triangular (D_3) symmetry instantaneously. Moreover, it has the spatio-temporal symmetry given by a 60° rotation combined with advance in time by one period T of the forcing, as in figure 2(e,g). In fact, this spatio-temporal symmetry was first suggested by the analysis below, and found to be consistent with the experimental observations. The third pattern in figure 2(c,f) is quite different: the dark lozenges in figure 2(f) represent the enlarged gaps between the hexagons in figure 2(c). The pattern is periodic in the medium-sized dashed hexagon in figure 2(f), so is invariant under translations τ_1^2 and $\tau_2^2 = \tau_1^4$. It is

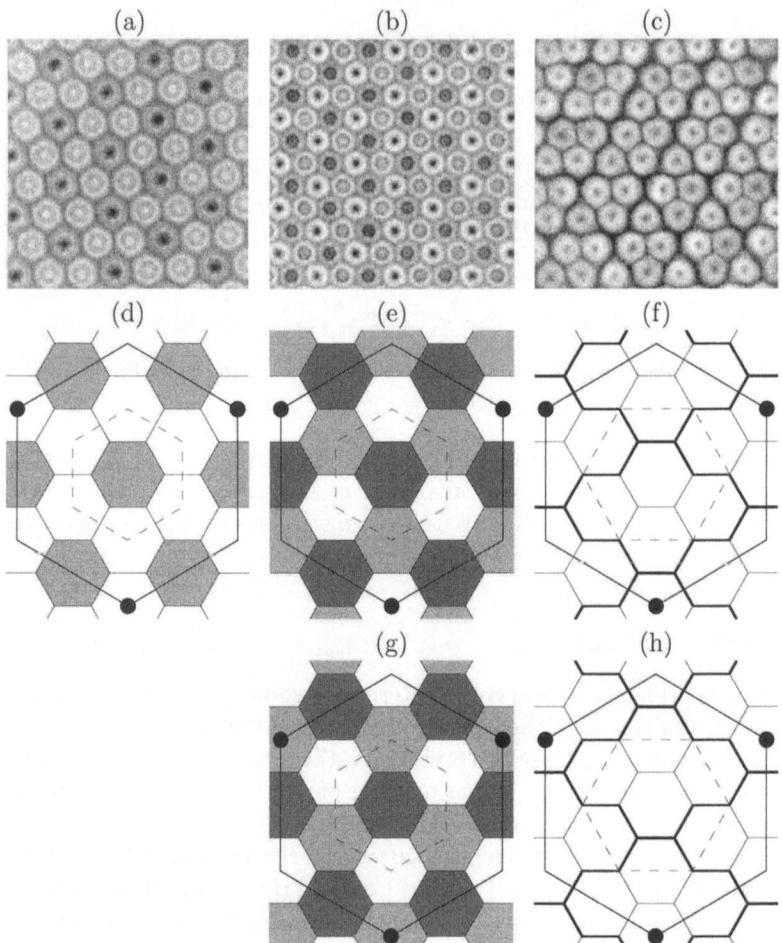

FIGURE 2. Experimental and idealised secondary patterns. (a-c) Experimental patterns, visualised from above. (d-f) Idealised versions of (a-c). (g-h) patterns (e-f) but seen one forcing period T later. The idealisations are all rotated by about 30° compared with the experimental pictures.

also invariant under the group of symmetries of a rectangle D_2, and possesses the spatio-temporal symmetry of the translation τ_2 combined with advance in time by one period T of the forcing, as in figure 2(f,h).

Using the information above, we write down the instantaneous (spatial) symmetry groups of the three patterns from figure 2(a-c) in terms of their generators:

$$\Sigma_a = \langle \kappa_x, \rho, \tau_1^3, \tau_1\tau_2 \rangle, \quad \Sigma_b = \langle \kappa_x, \rho^2, \tau_1^3, \tau_1\tau_2 \rangle, \quad \Sigma_c = \langle \kappa_x, \kappa_y\tau_2, \tau_1^2 \rangle. \quad (6)$$

These groups are of order 48, 24 and 12 respectively. For the full spatio-temporal symmetry groups, we would also include $\rho\tau_T$ in the generators of Σ_b, and $\tau_2\tau_T$

	a	b	c	d	e	f	g	h	i	j	k	l	m	n	o
	id	κ_x	κ_y	τ_1	τ_1^2	τ_1^3	$\kappa_x\tau_1$	$\kappa_x\tau_2$	$\kappa_x\tau_1^3$	$\kappa_y\tau_1^3$	ρ	ρ^2	ρ^3	$\rho^2\tau_1$	$\rho^3\tau_1^3$
	1	6	18	6	2	3	12	12	6	18	24	8	3	16	9
1	1	1	1	1	1	1	1	1	1	1	1	1	1	1	1
2	1	-1	-1	1	1	1	-1	-1	-1	-1	1	1	1	1	1
3	1	1	-1	1	1	1	1	1	1	-1	-1	1	-1	1	-1
4	1	-1	1	1	1	1	-1	-1	-1	1	-1	1	-1	1	-1
5	2	0	0	2	2	2	0	0	0	0	1	-1	-2	-1	-2
6	2	0	0	2	2	2	0	0	0	0	-1	-1	2	-1	2
7	2	2	0	-1	-1	2	-1	-1	2	0	0	2	0	-1	0
8	2	-2	0	-1	-1	2	1	1	-2	0	0	2	0	-1	0
9	3	1	1	-1	3	-1	-1	1	-1	-1	0	0	3	0	-1
10	3	-1	1	-1	3	-1	1	-1	1	-1	0	0	-3	0	1
11	3	-1	-1	-1	3	-1	1	-1	1	1	0	0	3	0	-1
12	3	1	-1	-1	3	-1	-1	1	-1	1	0	0	-3	0	1
13	4	0	0	-2	-2	4	0	0	0	0	0	-2	0	1	0
14	6	-2	0	1	-3	-2	-1	1	2	0	0	0	0	0	0
15	6	2	0	1	-3	-2	1	-1	-2	0	0	0	0	0	0

TABLE 1. Character table of the group Γ, taken from Tse *et al.*, with corrections. A representative element is shown on the second line for each conjugacy class (see also figure 3), and the number of elements in the class is on the third row. The next fifteen rows give the characters associated with each conjugacy class for each of the fifteen representations.

in the generators of Σ_c, but initially we will work with the spatial symmetry groups. The reason for this is that the instantaneous (spatial) symmetries can be determined reliably from a single experimental image, while extracting spatio-temporal symmetries from the experimental data is more involved.

Each of the three instabilities that generates the three different patterns will be associated with a set of marginally stable eigenfunctions; the new pattern, at least near onset, can be thought of as (approximately) a linear combination of these marginal eigenfunctions and the original hexagonal pattern. Which linear superpositions are consistent with the nonlinearity inherent in the pattern formation process is determined by our bifurcation analysis. The symmetries in Γ all leave the primary hexagonal pattern unchanged, so they must send marginal eigenfunctions onto linear combinations of marginal eigenfunctions, which induces an action on the amplitudes of these functions. In other words, if there are n marginal eigenfunctions $f_1, \ldots, f_n$, with n amplitudes $\mathbf{a} = (a_1, \ldots, a_n) \in \mathbb{R}^n$, each element $\gamma \in \Gamma$ sends $\mathbf{a}$ to $R_\gamma \mathbf{a}$, where the set of $n \times n$ orthogonal matrices R_γ forms a representation R_Γ of the group Γ. For subharmonic instabilities of the type of interest here, this will generically be an irreducible representation (irrep) [4]. Tse *et al.* [5] have computed all the irreps of the group Γ; the character table of these representations is reproduced in table 1. Recall that the character of a group element γ in a representation is the trace of the matrix R_γ, and that conjugate elements (which form a conjugacy class) have the same characters.

Once the representation associated with each of the three transitions is identified, we can write down the normal form, work out what other patterns are created in the same bifurcation, and compute stability of the patterns in terms of the normal form coefficients.

4. Method

The first task is to identify which representation is relevant for each bifurcation. Tse *et al.* [5] outlined a two-stage method to accomplish this. First, any symmetry element that is represented by the identity matrix in a particular representation must appear in the symmetry group of every branch of solutions created in a bifurcation with that representation. This can be used to eliminate from consideration any representation that has an element with character equal to the character of the identity that does not appear in the symmetry group of the observed pattern. Second, we make use of the trace formula from [4], which gives the dimension of the subspace of $\mathbb{R}^n$ that is fixed by a particular isotropy subgroup Σ of Γ with representation given by the matrices R_Γ:

$$\dim \text{fix}(\Sigma) = \frac{1}{|\Sigma|} \sum_{\sigma \in \Sigma} \text{Tr} \, R_\sigma, \tag{7}$$

where $|\Sigma|$ is the number of elements in Σ. Specifically, we use the trace formula to eliminate those representations for which the spatial symmetry group of the pattern fixes a zero-dimensional subspace (implying that the subgroup is not an isotropy subgroup); only the remaining representations need be examined in more detail.

We proceed by first counting the number of elements in each conjugacy class for each of the symmetry groups Σ_a, Σ_b and Σ_c. Figure 3 shows representative elements from each class and is helpful for this categorization. The result of this is: Σ_a contains:

$$a : 1, \, b : 6, \, c : 6, \, f : 3, \, i : 6, \, j : 6, \, k : 8, \, l : 8, \, m : 1, \, o : 3 \tag{8}$$

(that is, one element from class a, six from class b etc.); Σ_b contains:

$$a : 1, \, b : 6, \, f : 3, \, i : 6, \, l : 8; \tag{9}$$

and Σ_c contains:

$$a : 1, \, b : 1, \, c : 3, \, e : 2, \, h : 2, \, o : 3. \tag{10}$$

The element τ_1^2 does not appear in the symmetry groups of patterns (a) and (b), which eliminates representations 1–6 and 9–12 (since τ_1^2 is represented by the identity matrix in all these: see table 1). Similarly, $\tau_1 \tau_2$ in class f and ρ^3 do not appear in Σ_c, which eliminates representations 1–9, 11 and 13 from consideration for that bifurcation problem.

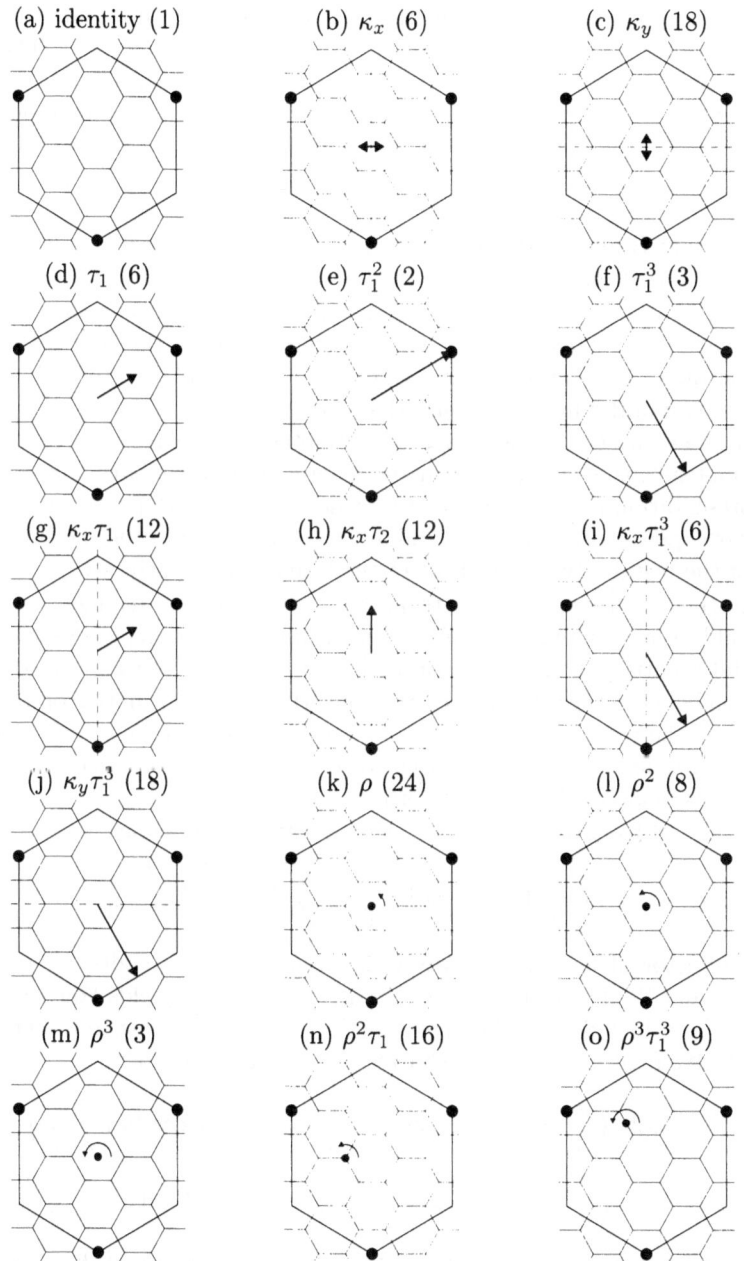

FIGURE 3. The 15 conjugacy classes of Γ. One element from and the number of elements in each class are indicated. The letters (a)–(o) correspond to the columns of table 1.

Next, by applying (7), we find that pattern (a) has a non-zero dimensional fixed point subspace only in representation 7, as does pattern (b). The spatial symmetry group of pattern (a) fixes a one-dimensional subspace, and that of pattern (b) fixes a two-dimensional subspace. Pattern (c) has a one-dimensional fixed point subspace in representations 10 and 12, and zero in other representations.

We are therefore faced with three different situations: the spatial symmetry group Σ_a fixes a one-dimensional subspace in representation 7, so we expect by the Equivariant Branching Lemma (see [4]) that such a pattern will generically be found in a bifurcation problem with that representation.

Pattern (b), on the other hand, has a spatial symmetry group that fixes a two-dimensional subspace. However, we must take into account that the pattern arises in a subharmonic (period-doubling) instability, and extend the groups Γ and Σ_b to the spatio-temporal symmetry groups that arise by including time translations. We may then show that the spatio-temporal symmetry group of pattern (b) fixes a one-dimensional subspace, and so also arises generically in a subharmonic bifurcation with representation 7. This is the same representation as with pattern (a), obtained for similar experimental parameter values. Extending to include the subharmonic nature of the instability does not affect the branching of pattern (a).

The third situation arises with pattern (c), which on symmetry arguments alone could be associated with either representation 10 or representation 12. Including information about the spatio-temporal symmetry of the pattern does not distinguish between these two representations. However, information on the Fourier transform of the pattern does allow a choice to be made between the two possibilities; in order to show this, we first need to work out which combinations of Fourier modes are associated with each pattern.

It is useful to have sample Fourier modes for the basic hexagonal pattern:

$$f_0(x,y) = \cos 2\pi \left(\frac{2x}{3}\right) + \cos 2\pi \left(-\frac{x}{3} + \frac{y}{\sqrt{3}}\right) + \cos 2\pi \left(-\frac{x}{3} - \frac{y}{\sqrt{3}}\right), \quad (11)$$

with wavevector of length $\frac{4\pi}{3}$, as well as sample Fourier modes for representations 7, 10 and 12. The method described by Tse et al. [5] yields Fourier functions that would be included in the eigenfunctions associated with representation 7; representative functions with the shortest wavevectors include:

$$f_1(x,y) = \cos 2\pi \left(\frac{x}{3} + \frac{y}{3\sqrt{3}}\right) + \cos 2\pi \left(\frac{x}{3} - \frac{y}{3\sqrt{3}}\right) + \cos 2\pi \left(\frac{2y}{3\sqrt{3}}\right) \quad (12)$$

$$f_2(x,y) = \sin 2\pi \left(\frac{x}{3} + \frac{y}{3\sqrt{3}}\right) + \sin 2\pi \left(-\frac{x}{3} + \frac{y}{3\sqrt{3}}\right) + \sin 2\pi \left(-\frac{2y}{3\sqrt{3}}\right), \quad (13)$$

which is made up of wavevectors of length equal to $\frac{1}{\sqrt{3}}$ of that of the basic hexagonal pattern. Eigenfunctions for representation 10 are made up of Fourier functions that include:

$$f_1 = \sin 2\pi \left(\frac{x}{6} + \frac{y}{2\sqrt{3}}\right) \quad f_2 = \sin 2\pi \left(\frac{-x}{6} + \frac{y}{2\sqrt{3}}\right) \quad f_3 = \sin 2\pi \left(\frac{-x}{3}\right), \quad (14)$$

with wavevector of length $\frac{1}{2}$ the fundamental; and representation 12 has:

$$f_1 = \sin 2\pi \left(\frac{x}{2} + \frac{-y}{2\sqrt{3}} \right) \quad f_2 = \sin 2\pi \left(\frac{x}{2} + \frac{y}{2\sqrt{3}} \right) \quad f_3 = \sin 2\pi \left(\frac{y}{\sqrt{3}} \right), \quad (15)$$

with wavevector of length $\frac{\sqrt{3}}{2}$ the fundamental. In each case, we have chosen the Fourier modes with the shortest wavevectors, as these are easiest to identify in an experimental Fourier transform.

The images of the Fourier transform of pattern (c) in [7] show that the mode created in the instability contains wavevectors that are a factor of 2 shorter than the shortest in the basic hexagonal pattern, which is consistent with representation 10 but not 12. In this way, information about the power spectrum of the pattern is necessary to supplement the arguments based entirely on symmetries and to distinguish between the two choices.

5. Normal forms

Using the functions specified above as a basis for representations 7 and 10, the matrices that generate the two relevant representations are, for representation 7:

$$R_{\kappa_x} = I_2, \quad R_\rho = \begin{bmatrix} 1 & 0 \\ 0 & -1 \end{bmatrix}, \quad R_{\tau_1} = \begin{bmatrix} -\frac{1}{2} & \frac{\sqrt{3}}{2} \\ -\frac{\sqrt{3}}{2} & -\frac{1}{2} \end{bmatrix}, \quad R_{\tau_2} = R_{\tau_1}^2, \quad R_{\tau\tau} = -I_2,$$

$$(16)$$

where I_n is the $n \times n$ identity matrix; and for representation 10:

$$R_{\kappa_x} = \begin{bmatrix} 0 & 1 & 0 \\ 1 & 0 & 0 \\ 0 & 0 & -1 \end{bmatrix}, \quad R_\rho = \begin{bmatrix} 0 & 0 & -1 \\ 1 & 0 & 0 \\ 0 & 1 & 0 \end{bmatrix}, \quad (17)$$

$$R_{\tau_1} = \begin{bmatrix} -1 & 0 & 0 \\ 0 & 1 & 0 \\ 0 & 0 & -1 \end{bmatrix}, \quad R_{\tau_2} = \begin{bmatrix} -1 & 0 & 0 \\ 0 & -1 & 0 \\ 0 & 0 & 1 \end{bmatrix}, \quad R_{\tau\tau} = -I_3. \quad (18)$$

The perturbation amplitude at time $j + 1$ times the forcing period, given the perturbation at time j, is given by $\mathbf{a}_{j+1} = \mathbf{f}(\mathbf{a}_j)$, where the equivariance condition amounts to $R_\gamma \mathbf{f}(\mathbf{a}) = \mathbf{f}(R_\gamma \mathbf{a})$ for all $\gamma \in \Gamma$. Using this, we can determine the relevant normal form associated with these two representations:

$$z_{j+1} = -(1 + \mu)z_j + P|z_j|^2 z_j + Q|z_j|^4 z_j + R\bar{z}^5 \quad (19)$$

for representation 7 (truncated at quintic order), where the two amplitudes of f_1 and f_2 in (12–13) are the real and imaginary parts of z, and P, Q and R are real constants. For representation 10 we truncate at cubic order and obtain:

$$a_{j+1} = -(1 + \mu)a_j + Pa_j^3 + Q(a_j^2 + b_j^2 + c_j^2)a_j, \quad (20)$$

$$b_{j+1} = -(1 + \mu)b_j + Pb_j^3 + Q(a_j^2 + b_j^2 + c_j^2)b_j, \quad (21)$$

$$c_{j+1} = -(1 + \mu)c_j + Pc_j^3 + Q(a_j^2 + b_j^2 + c_j^2)c_j, \quad (22)$$

where P and Q are (different) real constants. In these two sets of equations, μ represents the bifurcation parameter. The -1 Floquet multipliers at $\mu = 0$ arise because these are subharmonic bifurcations. In representation 7, equivariance with respect to $R_{\tau_T} = -I_2$ is a normal form symmetry, so even terms up to any order can be removed from (19) by coordinate transformations [9]. With representation 10, the matrix $-I_3 = R_\rho^3$ appears as a spatial symmetry, so the normal form symmetry is in fact exact, and every solution branch has the spatio-temporal symmetry $\tau_T \rho^3$, a rotation by $180°$ followed by time-translation by one period.

The patterns are neutrally stable with respect to translations in the two horizontal directions, and so also have two Floquet multipliers equal to 1 associated with translation modes. We have neglected these as all the patterns we find are pinned by reflection symmetries that prohibit drifting.

The final stages are to determine the solutions that are created in each of these bifurcations, their symmetry and stability properties, and to compare these with experimental observations.

The first normal form (19) generically has two types of period-two points, found by solving $f(z) = -z$:

$$z_a = \sqrt{\frac{\mu}{P} - 2\mu^2 \frac{Q+R}{P^3}}, \qquad z_b = i\sqrt{\frac{\mu}{P} - 2\mu^2 \frac{Q-R}{P^3}}. \tag{23}$$

The first of these has exactly the symmetry group Σ_a of pattern (a), with no spatio-temporal symmetries, while the second has exactly the spatial symmetry group Σ_b of pattern (b), as well as spatio-temporal symmetries generated by $\rho\tau_T$. Reconstructions of these two are shown in figure 4(a) for pattern (a) and figure 4(b,c) for pattern (b), using the Fourier functions from above. Linearising the normal form about these two period-two points readily yields stability information: if $P > 0$, then both patterns are supercritical but only one is stable, while if $P < 0$, both are subcritical and neither is stable.

The second normal form (20–22) generically has three types of period-two points (a, b, c):

$$\sqrt{\frac{\mu}{P+Q}}\begin{pmatrix}1\\0\\0\end{pmatrix}, \quad \sqrt{\frac{\mu}{P+2Q}}\begin{pmatrix}1\\1\\0\end{pmatrix}, \quad \sqrt{\frac{\mu}{P+3Q}}\begin{pmatrix}1\\1\\1\end{pmatrix}. \tag{24}$$

The middle branch has the spatio-temporal symmetries of pattern (c), with 12 elements in the spatial part of the symmetry group ($\Sigma_c = \langle \kappa_x, \kappa_y\tau_2, \tau_1^2 \rangle$). Figure 5(a,b) illustrates this pattern (cf. figure 2c,f,h). For comparison, the pattern that would have been obtained with modes from representation 12 is in figure 5(c,d): the symmetry group is the same, but the appearance of the pattern does not match the experimental observation. The first branch has a 24 element spatial symmetry group $\langle \rho^3\tau_1, \kappa_x\rho\tau_1^5\tau_2, \tau_1^2 \rangle$ (figure 5e,f), and the third branch has an 18 element group $\langle \kappa_y\tau_2, \kappa_x\rho^5, \tau_1^2 \rangle$ (figure 5g,h). The three patterns also have the spatio-temporal symmetry $\rho^3\tau_T$ (since $R_\rho^3 = -I_3$), so ρ^3 will appear in the symmetry group of the time-average of each of the patterns, as discussed in [5].

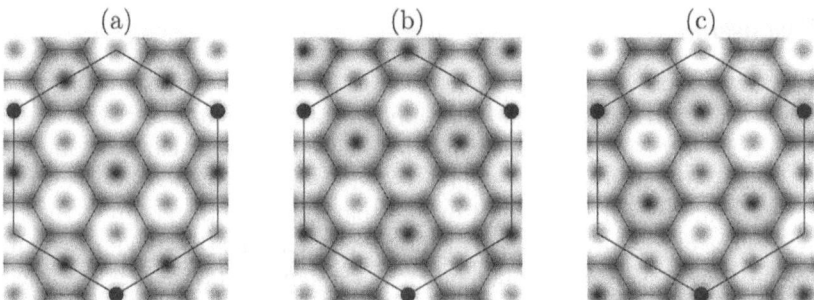

FIGURE 4. Reconstructed patterns from the two solutions that arise in representation 7, using the Fourier functions (12–13) added to a function of the form of (11). (a) has the spatial symmetries of pattern (a) and no spatio-temporal symmetries (cf. 2a,d); (b) has the symmetry properties of pattern (b) (c is one period T later; cf. figure 2b,e,g)

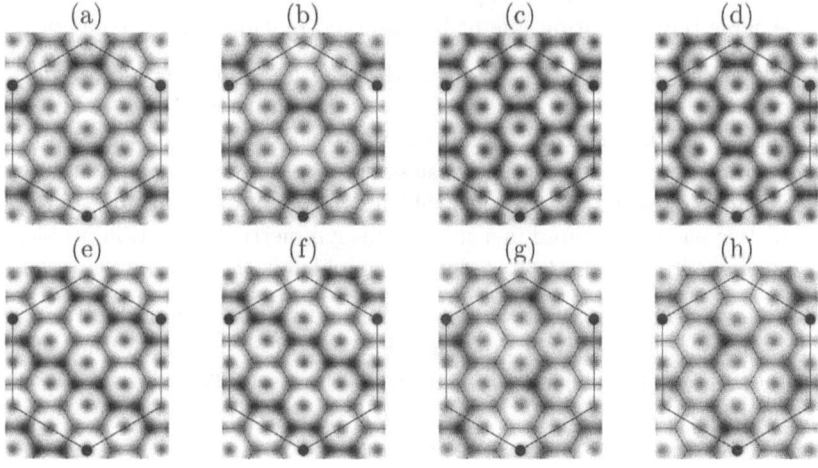

FIGURE 5. Reconstructed patterns from irreps 10 and 12: (a,b) irrep 10, with amplitudes $(a, b, c) = (1, 1, 0)$ (cf. figure 2c,f,h); (c,d) irrep 12: same amplitudes and same symmetries as (a,b); (e,f) irrep 10, with amplitudes $(a, b, c) = (1, 0, 0)$; (g,h) irrep 10, with amplitudes $(a, b, c) = (1, 1, 1)$.

The first branch has Floquet multipliers $-1 + 2\mu$ and $-1 - \frac{P}{P+Q}\mu$ (twice); the second branch $-1 + 2\mu$, $-1 - \frac{P}{P+2Q}\mu$ and $-1 + \frac{2P}{P+2Q}\mu$; and the third branch $-1 + 2\mu$ and $-1 + \frac{2P}{P+3Q}\mu$ (twice). As a result, if $P + Q > 0$ and $P + 3Q > 0$, then all branches bifurcate supercritically, and either the first branch will be stable (when $P < 0$) or the last will be stable (when $P > 0$). If any branch bifurcates subcritically, none are stable. The middle branch, which is the one corresponding to the experimentally observed pattern (c), is always unstable at onset.

6. Discussion

Using the symmetry-based approach of Tse *et al.* [5], we have analysed three experimentally observed spatial period-multiplying transitions from an initial hexagonal pattern. The three patterns illustrate three situations that can arise in this kind of analysis. Pattern (a) was straight-forward, in that a single representation of Γ had a one-dimensional space fixed by the spatial symmetry group of the pattern. The existence of a solution branch of the form of pattern (b) could also be inferred using the Equivariant Branching Lemma, though in this case it was necessary to include the temporal symmetry associated with period-doubling bifurcation. Specifically, the spatial symmetries selected a two-dimensional fixed point space which was further reduced to a one-dimensional fixed point space when spatio-temporal symmetries were taken into account. Experimentally, these two patterns were found for the same fluid parameters and same $2\omega : 3\omega$ forcing function but for different frequencies ω: $\omega = 25\,\mathrm{Hz}$ for (a) and $\omega = 35\,\mathrm{Hz}$ for (b). This suggests that the transition between these patterns, which arise for instabilities associated with the same representation, might be observed by tuning the frequency ω.

Pattern (c), on the other hand, had a spatial symmetry group that fixed one-dimensional subspaces in two different representations, and we appealed to the measured power spectrum of the pattern to choose between the two possibilities. In this situation, symmetry considerations alone were not enough. Similar situations arise in other bifurcation problems, for example, knowing that a stable axisymmetric pattern is found in a spherically symmetric bifurcation problem does not provide enough information to determine which is the relevant representation.

The experimentally observed transition between hexagons and pattern (c) occurs by means of a propagating front that separates domains of hexagons and the secondary pattern. The front is initiated at the lateral boundaries of the system and emanates radially inward. There is little if any hysteresis, and the reverse transition also occurs via the same scenario. The occurrence of a front in this transition suggests bistability of the hexagonal pattern and pattern (c). This is certainly consistent with the theoretical prediction that pattern (c) is unstable at small amplitude, that is, at onset. However, we have not explored the possible stabilization mechanisms for pattern (c).

It is worth emphasizing that an understanding of group representation theory is useful in classifying and analysing secondary instabilities of patterns, not only in the Faraday wave experiment as described here, but also in convection and other pattern formation problems (see [10]). It is also worth mentioning that the examples studied here indicate that spatio-temporal symmetries readily arise in secondary subharmonic instabilities, and that careful experimental characterization of these, either by still images taken one forcing period apart or by time-averaging over two forcing periods, can be helpful. Subsequent instabilities of patterns that have spatio-temporal symmetries can be analysed using methods described in [11, 12].

The approach outlined in [5] and here is useful for taking an experimental observation of a secondary transition and casting it into its equivariant bifurcation theory context, but it does not predict which transitions should be expected in an experiment. However, in these two-frequency Faraday wave experiments, three-wave interactions of the type described in [13] may select a third wavevector that could appear in the secondary transition. Each of the representations in the problem under consideration is associated with a set of wavevectors, providing a possible mechanism for selecting between possibilities.

Acknowledgements

This paper builds on earlier published results obtained with Dawn Tse, Rebecca Hoyle and Hagai Arbell. AMR is grateful for support from the EPSRC. The research of MS is supported in part by NSF grant DMS-9972059 and NASA grant NAG3-2364. JF is grateful for support from the Israel Academy of Science (grant 203/99).

References

[1] Edwards, W.S. & Fauve, S., *Patterns and quasi-patterns in the Faraday experiment*, J. Fluid Mech., **278** (1994), 123–148.

[2] Kudrolli, A. & Gollub, J.P., *Patterns and spatiotemporal chaos in parametrically forced surface waves: a systematic survey at large aspect ratio*, Physica, **97D** (1996), 133–154.

[3] Müller, H.W., Friedrich, R. & Papathanassiou, D. (1998) Theoretical and experimental investigations of the Faraday instability. In *Evolution of Spontaneous Structures in Dissipative Continuous Systems* (ed. F.H. Busse & S.C. Müller), 230–265. Springer: Berlin

[4] Golubitsky, M., Stewart, I. & Schaeffer, D.G. (1988) *Singularities and Groups in Bifurcation Theory. Volume II*. Springer: New York.

[5] Tse, D.P., Rucklidge, A.M., Hoyle, R.B. & Silber, M., *Spatial period-multiplying instabilities of hexagonal Faraday waves*, Physica, **146D** (2000), 367–387.

[6] Kudrolli, A., Pier, B. & Gollub, J.P., *Superlattice patterns in surface waves*, Physica, **123D** (1998), 99–111.

[7] Arbell, H. & Fineberg, J., *Spatial and temporal dynamics of two interacting modes in parametrically driven surface waves*, Phys. Rev. Lett., **81** (1998), 4384–4387.

[8] Lioubashevski, O., Arbell, H. & Fineberg, J., *Dissipative solitary states in driven surface waves*, Phys. Rev. Lett., **76** (1996), 3959–3962.

[9] Elphick, C., Tirapegui, E., Brachet, M.E., Coullet, P. & Iooss, G., *A simple global characterization for normal forms of singular vector fields*, Physica, **29D** (1987), 95–127.

[10] Rucklidge, A.M., Weiss, N.O., Brownjohn, D.P., Matthews, P.C. & Proctor, M.R.E., *Compressible magnetoconvection in three dimensions: pattern formation in a strongly stratified layer*, J. Fluid Mech., **419** (2000), 283–323.

[11] Rucklidge, A.M. & Silber, M., *Bifurcations of periodic orbits with spatio-temporal symmetries,* Nonlinearity, **11** (1998), 1435–1455.

[12] Lamb, J.S.W. & Melbourne, I. (1999) Bifurcation from periodic solutions with spatiotemporal symmetry. In *Pattern Formation in Continuous and Coupled Systems* (ed. M. Golubitsky, D. Luss & S.H. Strogatz), 175–191. Springer: New York

[13] Silber, M., Topaz, C.M. & Skeldon, A.C., *Two-frequency forced Faraday waves: weakly damped modes and patterns selection,* Physica, **143D** (2000), 205–225.

A.M. Rucklidge
Department of Applied Mathematics
University of Leeds
Leeds LS2 9JT UK
e-mail: `A.M.Rucklidge@leeds.ac.uk`

M. Silber
Department of Engineering Sciences and Applied Mathematics,
Northwestern University
Evanston IL 60208 USA
e-mail: `m-silber@northwestern.edu`

J. Fineberg
The Racah Institute of Physics
The Hebrew University of Jerusalem
Givat Ram
Jerusalem 91904 Israel
e-mail: `jay@vms.huji.ac.il`

Contributed Papers

Trends in Mathematics:
Bifurcations, Symmetry and Patterns, 117–122
© 2003 Birkhäuser Verlag Basel/Switzerland

Spatially Resonant Interactions in Annular Convection

Arantxa Alonso, Marta Net and Juan Sánchez

Abstract. Different types of steady columnar patterns in an annular container with a fixed value of the radius ratio are analyzed for a low Prandtl number Boussinesq fluid. The stability of these convection patterns as well as the spatial interaction between them resulting in the formation of mixed modes are numerically investigated by considering the original nonlinear set of Navier-Stokes equations. A detailed picture of the nonlinear dynamics before temporal chaotic patterns set in is presented and understood in terms of symmetry-breaking bifurcations in an $\mathbf{O}(2)$-symmetric system. Special attention is paid to the strong spatial 1:2 resonance of the initially unstable modes with wavenumbers $n=2$ and $n=4$, which leads to bistability in the system.

1. Introduction

In the present paper we investigate the process of pattern selection and mode interaction in the context of two-dimensional thermal convection. We analyze convection in a rotating annulus with gravity radially inwards and outwards heating, restricting our attention to exactly two-dimensional solutions. These solutions, which form columns parallel to the axis of rotation, are allowed when stress-free boundary conditions on the lids of the annulus are considered, and are the preferred modes at the onset of convection for large enough rotation rates [1].

When the two-dimensional governing equations are considered the symmetry of the system is $\mathbf{O}(2)$ even in the rotating case. Although rotation breaks the reflection symmetry in vertical planes containing the axis, for the columnar solution the Coriolis term can be written as a gradient and introduced in the pressure term. Rotation drops from the equations and they retrieve the reflection invariance. Thus, columnar convection in a rotating annulus provides a simple fluid dynamics system with $\mathbf{O}(2)$ symmetry, which exhibits a rich variety of stationary and spatio-temporal patterns [2], [3], [4].

In a previous work [2], we studied the transition route to chaos that the stable columnar solution with wavenumber $n=3$ undergoes for a low value of the Prandtl number, $\sigma=0.025$, and a fixed value of the radius ratio, $\eta=0.3$. The steady columns give rise to spatially periodic and non-periodic direction reversing travelling waves, which become chaotic through a Neimark-Sacker subcritical bifurcation. In order to complete the analysis, we will now focus on the unstable branches that bifurcate from the conduction state. We will see that there is a strong spatial interaction between the two initially unstable modes with wavenumbers $n=2$ and $n=4$ and that some aspects of the behaviour we find, such as the presence of a mixed mode, wavenumber gaps and travelling waves, are predicted by the normal form equations for the 1:2 spatial resonance with $O(2)$ symmetry [5], [6].

2. Formulation of the problem

We consider the problem of nonlinear convection in a cylindrical annulus with radius ratio $\eta = r_i/r_o$, where r_i and r_o are the inner and outer radii, rotating about its axis of symmetry, filled with a Boussinesq fluid of thermal diffusivity κ, thermal expansion coefficient α and kinematic viscosity ν. The inner and outer rigid sidewalls are maintained at constant temperatures T_i and T_o, with $T_i > T_o$, and the gravitational acceleration is taken radially inwards, $\mathbf{g} = -g\hat{\mathbf{e}}_r$, and is assumed to be constant.

There exists a basic conduction state in which heat is radially transferred towards the outer cylinder by thermal conduction and which shares the full symmetries of the system. The stability of this state is determined by the Navier-Stokes, continuity and heat equations. When horizontal stress-free lids are considered, the linear stability analysis shows that there is always a moderate rotation rate above which steady exactly two-dimensional columns parallel to the axis of the annulus are the preferred solutions at the onset of convection [1]. These solutions are characterized by a fixed azimuthal wavenumber, n, which is imposed by the chosen radius ratio.

In order to obtain the nonlinear steady columnar solutions and to analyze their stability with respect to axial independent disturbances as any parameter of interest is varied, we have developed a continuation code [7]. To solve the equations we have used a technique based on velocity potentials. In this formulation the velocity field is written as $\mathbf{u} = f\hat{\mathbf{e}}_\theta + \nabla \times \Psi\hat{\mathbf{e}}_z$, where $\Psi=\Psi(r,\theta)$ is a function which has zero azimuthal average. We are considering an explicit equation for $f = f(r)$, which is the simplest way of including the possibility of generating a mean mass flow in the azimuthal direction (average of the azimuthal velocity in the radial direction). The resulting nonlinear equations in nondimensional form are

$$\partial_t f = \sigma\nabla_-^2 f + P_\theta\left[\nabla_h^2\Psi\left(\frac{1}{r}\partial_\theta\Psi\right)\right], \tag{1a}$$

$$\partial_t\nabla_h^2\Psi = \sigma\nabla_h^4\Psi + (1 - P_\theta)\frac{\sigma Ra}{r}\partial_\theta\Theta + (1 - P_\theta)J(\Psi,\nabla_h^2\Psi) +$$

$$+ \nabla_-^2 f\left(\frac{1}{r}\partial_\theta\Psi\right) - f\left(\frac{1}{r}\partial_\theta\nabla_h^2\Psi\right), \tag{1b}$$

$$\partial_t\Theta = \nabla_h^2\Theta - \frac{1}{r^2\ln\eta}\partial_\theta\Psi + J(\Psi,\Theta) - f\left(\frac{1}{r}\partial_\theta\Theta\right), \tag{1c}$$

where Θ denotes the departure of the temperature from its conduction profile and $\nabla_-^2 = \partial_r(\partial_r + 1/r)$. P_θ is the projection operator that extracts the zero-azimuthal mode in a Fourier expansion, J is the jacobian in cylindrical coordinates and Ra and σ are the Rayleigh and Prandtl numbers. The variables have been expanded in terms of Chebyshev polynomials and Fourier expansions and no-slip and perfectly conducting boundary conditions on the lateral walls have been considered.

3. Nonlinear steady columnar solutions: results and discussion

In this section we describe the results for $\eta=0.3$, $\sigma=0.025$ and increasing Rayleigh number. We obtain the steady columnar patterns that bifurcate from the conduction state and analyze their stability.

The results are summarized in figure 1. In the upper part of the figure, the bifurcation diagram shows the branches of columnar solutions with basic wavenumbers $n=3,2,4$ ($N3$, $N2$ and $N4$ branches, respectively). In the diagram, we are plotting an amplitude of the dominant mode in each case. The conduction state becomes unstable to columns with wavenumber $n=3$ at $Ra_c^1=1799$ (point 1 in the bifurcation diagram), in agreement with the linear stability analysis. For slightly larger Rayleigh numbers, the conduction state is also unstable to modes with wavenumber $n=2$ (at $Ra_c^2=1995$) and $n=4$ (at $Ra_c^3=2254$). The new nonaxisymmetric solutions break the rotation symmetry, $\mathbf{R}_\theta$, of the basic state, but maintain the reflection symmetry, $\mathbf{R}_1$, with respect to appropriate vertical planes $\theta = \theta_0$ and the invariance under $2\pi/n$-rotations, $\mathbf{R}_{2\pi/n}$. The group of symmetry of the new solutions is $\mathbf{D}_n$. Thus, bifurcations from the conduction state are symmetry-breaking steady-state bifurcations in which multiplicity two eigenvalues cross the imaginary axis.

Whereas solutions along the $N3$ and $N4$ branches are pure modes, in which only the basic wavenumbers and their harmonics are nonzero, the $N2$ is a mixed-mode branch. There is a strong spatial interaction between the $n=2$ and $n=4$ modes which produces a change in the structure of the solution along the $N2$ branch. To illustrate the physical nature of these solutions, the lower part of figure 1 shows the temperature contour plots at different Rayleigh numbers. As the Rayleigh number increases, the contribution of the $n=4$ mode becomes more and more important, while the $n=2$ contribution diminishes until vanishing. The initial two pairs of rolls become a $n=4$ solution.

A stability analysis of the mixed-mode solutions shows that there are several bifurcations in the $N2$ branch. The new branches have been included in figure 2. Bifurcation points 4 ($Ra_c^4=2362$), 6 ($Ra_c^6=2509$) and 7 ($Ra_c^7=2712$) correspond to subharmonic steady-state bifurcations. The solutions in these new branches, which are displayed in the right-hand side of figure 2, still keep the reflection symmetry between columns, but now there is a contribution of all the wavenumbers. Their

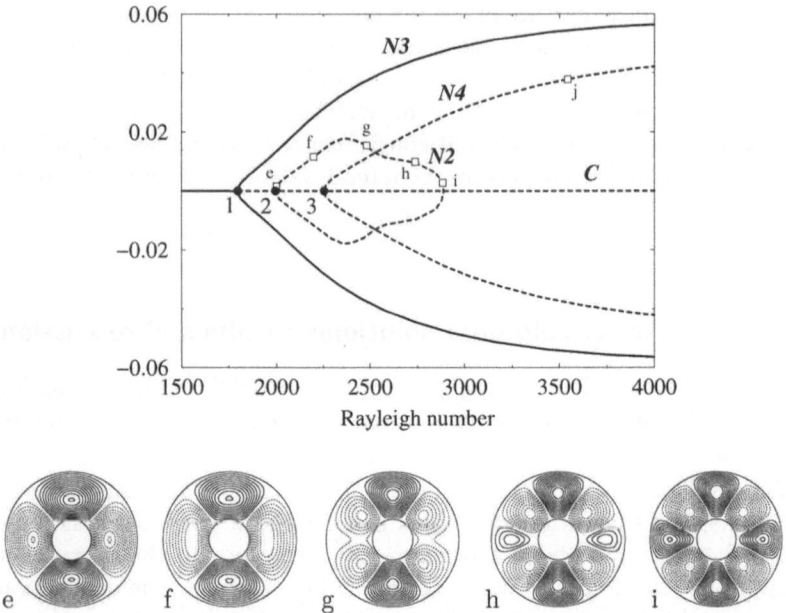

FIGURE 1. (top) Bifurcation diagram showing branches of columnar modes with wavenumbers $n=3$, $n=2$ and $n=4$ ($N3$, $N2$ and $N4$
branches, respectively). They are born at $Ra_c^1=1799$ (1), $Ra_c^2=1995$
(2) and $Ra_c^3=2254$ (3). (bottom) Temperature contour plots showing the evolution of the columns on the $N2$ branch with increasing
Rayleigh number. They correspond to points e ($Ra=2000$), f ($Ra=2198$),
g ($Ra=2500$), h ($Ra=2711$) and i ($Ra=2875$) in the diagram.

group of symmetry is $\mathbf{Z}_2$. The bifurcation identified in point 5 ($Ra_c^5=2478$) corresponds to a steady-state instability that keeps the wavenumber of the main solution, $n=2$, but in which the mean flow becomes nonzero. According to bifurcation
theory [8], a steady-state bifurcation that breaks the reflection symmetry keeping
the rotational invariance would give rise to travelling waves with zero phase speed
at the bifurcation point. Nevertheless, we have not followed this time-dependent
branch which, in our case, is unstable. Finally, two subsequent bifurcations very
close to each other take place in the neighbourhood of point 8. In the first one
($Ra_c^8=2887.5$), one of the two positive eigenvalues of the solution is stabilized
through a subharmonic steady-state bifurcation. In the second one ($Ra_c^{8'}=2888.9$)
the amplitude of the $n=2$ mode vanishes. The $N2$ branch joins the $N4$ branch, and
columns with wavenumber $n=2$ cease to exist. This is a bifurcation from the $N4$
branch, which takes place after a bifurcation in $Ra_c^9=2851$ in which an eigenvalue
with multiplicity two gains stability.

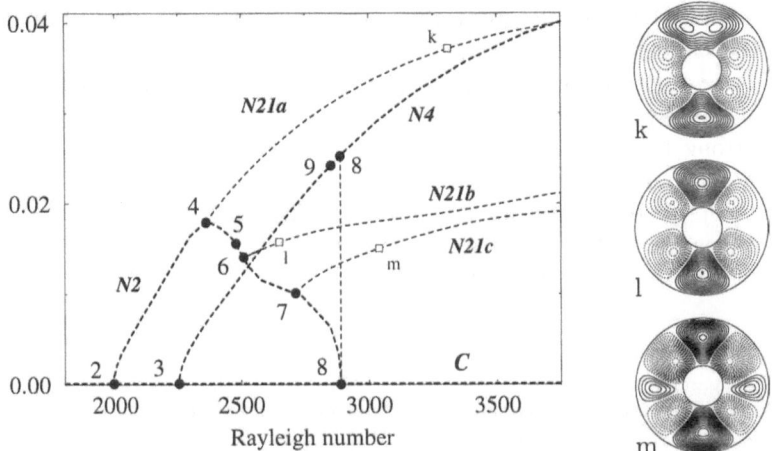

FIGURE 2. (left) Detail of the steady-state bifurcations on the $N2$ and $N4$ branches, which take place at $Ra_c^4=2362$ (4), $Ra_c^5=2478$ (5), $Ra_c^6=2509$ (6), $Ra_c^7=2712$ (7) and $Ra_c^8=2888$ (8) on the $N2$ branch and at $Ra_c^9=2851$ (9) and $Ra_c^8=2888$ (8) on the $N4$ branch. (right) Temperature contour plots showing the structure of the solutions in the $N21a$, $N21b$ and $N21c$ branches corresponding to points k ($Ra=3308$), l ($Ra=2649$) and m ($Ra=3040$) in the bifurcation diagram.

All the steady patterns above described except for the $n=3$ column are unstable. However, by extending further the $N4$ branch, a bifurcation that stabilizes the $n=4$ mode by shedding a new unstable $n=2$ branch takes place at $Ra_c^{10}=4779$. As a result, for Rayleigh numbers larger than Ra_c^{10} at least two stable solutions coexist: steady columns with wavenumber $n=4$ and direction reversing travelling waves with wavenumber $n=3$.

The spatial interaction between the modes with wavenumbers $n=2$ and $n=4$ that we have found is an example of an 1:2 resonance, in which modes with wavenumbers n and $2n$ in the periodic direction interact nonlinearly. The 1:2 resonance with $\mathbf{O}(2)$ symmetry was first studied by Dangelmayr [5] and some aspects of the dynamics predicted by the normal form equations are reproduced here. First, the presence of wavenumber gaps in which no steady solutions with a given wavenumber exist is a typical feature of this resonance. This is what happens in the range of Rayleigh numbers $2888 < Ra < 4779$, where the $n=2$ solution disappears. The existence of travelling waves bifurcating from the n-mode, which correspond to the bifurcation point at $Ra_c^5=2478$ in our case, is also an identity sign of this resonance.

Dynamics dominated by the strong spatial 1:2 resonance are expected in systems without midplane layer symmetry. In the case of annular convection this symmetry is broken by curvature, while in two-dimensional Rayleigh-Bénard convection, the same effect can be achieved by considering different boundary conditions at top and bottom or by including non-Boussinesq terms [9]. For instance, the analysis of a long-wave model for two-dimensional convection in a plane layer shows that the strong 1:2 resonance is dominant when asymmetric boundary conditions are considered and a behaviour similar to the one we find in annular convection is described [10]. In contrast, in Rayleigh-Bénard convection with symmetric boundary conditions the leading order resonant term in the 1:2 interaction is of higher order than that in the 1:3 interaction [10], [11], [12].

References

[1] A. Alonso, M. Net and E. Knobloch, *On the transition to columnar convection*, Phys. Fluids, **7 (5)** (1995), 935–940.

[2] A. Alonso, J. Sánchez and M. Net *Transition to temporal chaos in an O(2)-symmetric convective system for low Prandtl numbers*, Prog. Theor. Phys. Suppl., **139** (2000), 315–324.

[3] K. Zhang, C.X. Chen and G.T. Greed, *Finite-amplitude convection rolls with integer wavenumbers*, Geophys. Astrophys. Fluid Dynamics, **90** (1999), 265–296.

[4] W.T. Langford and D.D. Rusu, *Pattern formation in annular convection*, Physica A, **261** (1998), 188–203.

[5] G. Dangelmayr, *Steady-state mode interactions in the presence of O(2)-symmetry*, Dynamics and Stability of Systems, **1** (1986), 159–185.

[6] M. R. E. Proctor and C. A. Jones, *The interaction of two spatially resonant patterns in thermal convection. I. Exact 1:2 resonance*, J. Fluid Mech., **188** (1988), 301–335.

[7] J. Sánchez, *Simulación numérica en flujos confinados: estructuras preturbulentas*, PhD thesis, Universidad de Barcelona, (1984).

[8] E. Knobloch, *Symmetry and instability in rotating hydrodynamic and magnetohydrodynamic flows*, Phys. Fuids, **8 (6)** (1996), 1446–1454.

[9] G. Manogg and P. Metzener, *Strong resonance in two-dimensional non-Boussinesq convection*, Phys. Fluids, **6 (9)** (1994), 2944–2955.

[10] S. M. Cox, *Mode interactions in Rayleigh-Bénard convection*, Physica D, **95** (1996), 50–61.

[11] J. Mizushima and K. Fujimura, *Higher harmonic resonance of two-dimensional disturbances in Rayleigh-Bénard convection*, J. Fluid Mech., **234** (1992), 651–667.

[12] J. Prat, I. Mercader and E. Knobloch, *Resonant mode interactions in Rayleigh-Bénard convection*, Phys. Rev. E, **58 (3)** (1998), 3145–3156.

Arantxa Alonso, Marta Net and Juan Sánchez
Departament de Física Aplicada
Universitat Politècnica de Catalunya
Campus Nord, Mòdul B4
08034 Barcelona, Spain
e-mail: arantxa@fa.upc.es

Trends in Mathematics:
Bifurcations, Symmetry and Patterns, 123–127

Hopf Bifurcations on Cubic Lattices

T.K. Callahan

Abstract. We analyze three-dimensional pattern forming Hopf bifurcations with the spatial periodicity of the face-centered (FCC) and body-centered (BCC) cubic lattices. This is an equivariant bifurcation with spatial symmetry $\Gamma = T^3 \dotplus O \oplus \mathbb{Z}_2$. By extending the group to a larger, wreath product group we can use the method of [5] to find all solution branches guaranteed by group theory to be primary. This work is an extension of that done for the steady state FCC and BCC bifurcations in [3, 4].

1. Introduction

Consider a pattern forming system that is homogeneous and isotropic in all three spatial directions, such as a reaction–diffusion system. It has a trivial, homogeneous solution with the symmetry of the Euclidean group $E(3)$. Suppose that as a bifurcation parameter λ increases through a critical value λ_c this state loses stability *via* a Hopf bifurcation to modes with nonzero wavenumber k_c and frequency ω. As done in [8], we restrict our attention to solutions that have the periodicity of some lattice so that a center manifold reduction can be performed. There are three fundamental lattices based on the cubic lattice: the simple (SC), face centered (FCC) and body centered (BCC) cubic. The steady state bifurcations on these three lattices were analyzed in [3, 4] and the Hopf bifurcation on the SC lattice was analyzed in [6]. We here study the Hopf bifurcation on the FCC and BCC lattices. Although motivated by reaction–diffusion problems, our results are model-independent, and are applicable to other pattern forming problems near onset.

2. Hopf bifurcations on the BCC lattice

We start with the twelve wavevectors $\pm \vec{k}_i$, $i = 1, \ldots, 6$, at the midponts of the edges of a cube, given by

$$\vec{k}_1 = \frac{k_c}{\sqrt{2}}(1,1,0) \qquad \vec{k}_2 = \frac{k_c}{\sqrt{2}}(0,1,1) \qquad \vec{k}_3 = \frac{k_c}{\sqrt{2}}(1,0,1)$$

$$\vec{k}_4 = \frac{k_c}{\sqrt{2}}(1,-1,0) \qquad \vec{k}_5 = \frac{k_c}{\sqrt{2}}(0,1,-1) \qquad \vec{k}_6 = \frac{k_c}{\sqrt{2}}(-1,0,1)$$

Let z_j (w_j) be the amplitude of a wave travelling in the $+\vec{k}_j$ $(-\vec{k}_j)$ direction. A real scalar field with the periodicity of the BCC lattice can, to lowest order, be put into the form

$$X(\vec{x}, t) = \sum_{j=1}^{4} \left[z_j(t)e^{i\vec{k}_j \cdot \vec{x} - i\omega t} + w_j(t)e^{-i\vec{k}_j \cdot \vec{x} - i\omega t} \right] + \text{c.c.} + \text{h.o.t.},$$

where by the Center Manifold Theorem the higher order spatial harmonics are quadratically small. We write the amplitudes as a vector

$$\mathbf{z} = (z_1, w_1; z_2, w_2; z_3, w_3; z_4, w_4; z_5, w_5; z_6, w_6).$$

After center manifold reduction we have twelve complex amplitude equations, which can be written in the form $\dot{\mathbf{z}} = \mathbf{f}(\mathbf{z})$.

The system has the spatial symmetry group $\Gamma = T^3 \dotplus \mathbb{O} \oplus \mathbb{Z}_2$, where T^3 is the three-torus of translations $\tau_{\vec{\alpha}}$ and $\hat{c} \in \mathbb{Z}_2$ represents inversion through the origin. These act on the amplitudes by

$$\tau_{\vec{\alpha}} \cdot z_j = e^{i\vec{k}_j \cdot \vec{\alpha}} z_j, \qquad \tau_{\vec{\alpha}} \cdot w_j = e^{-i\vec{k}_j \cdot \vec{\alpha}} w_j, \qquad \hat{c} \cdot z_j = w_j, \qquad \hat{c} \cdot w_j = z_j.$$

$\mathbb{O}$ is a representation of the octahedral group of orientation preserving symmetries of the cube and acts as a permutation of the $\vec{k}_j$'s. We have three independent translations and can thus choose at most three of the phases of the z_j's.

As a Hopf bifurcation, the system must be *equivariant* under $\Gamma_1 = \Gamma \times S^1$, i.e.,

$$\gamma \cdot \mathbf{f}(\mathbf{z}) = \mathbf{f}(\gamma \cdot \mathbf{z}), \quad \forall \gamma \in \Gamma_1,$$

where S^1 is a normal form symmetry corresponding to the periodicity in time [8]. That is, $\hat{\psi} \in S^1$ acts by the phase shift $\hat{\psi} \cdot \mathbf{z} = e^{i\psi}\mathbf{z}$. Every solution $\mathbf{z}$ has an *isotropy subgroup*

$$\Sigma(\mathbf{z}) = \left\{ \gamma \in \Gamma_1 \mid \gamma \cdot \mathbf{z} = \mathbf{z} \right\},$$

and each isotropy subgroup Σ has a *fixed point subspace*

$$\text{Fix}(\Sigma) = \left\{ \mathbf{z} \in \mathbb{C}^{12} \mid \sigma \cdot \mathbf{z} = \mathbf{z}, \quad \forall \sigma \in \Sigma \right\},$$

which is an invariant subspace under the flow of any Γ_1-equivariant system. For any solution $\mathbf{z}$ with isotropy subgroup Σ and any group element $\gamma \in \Gamma$, there is another solution $\gamma \cdot \mathbf{z}$ with conjugate isotropy subgroup $\gamma \Sigma \gamma^{-1}$. We consider conjugate solutions and isotropy subgroups to be equivalent.

By the *Equivariant Hopf Theorem* [8], for every isotropy subgroup $\Sigma \in \Gamma_1$ with two-dimensional fixed point subspace (called a $\mathbb{C}$-*axial subgroup*) there is at onset a branch of periodic solutions with isotropy Σ and frequency close to ω. We want to find all the $\mathbb{C}$-axial subgroups of $\Gamma_1 = \Gamma \times S^1$.

We do this by extending the group Γ_1 to a larger group Γ_2 that is easier to solve. To that end we first group the twelve amplitudes in this fashion:

$$\mathbf{z} = (z_1, w_1; z_4, w_4 \mid z_2, w_2; z_5, w_5 \mid z_3, w_3; z_6; w_6). \qquad (*)$$

We consider these to be three subsystems, each with two orthogonal sets of wave-vectors $\vec{k}_i, \vec{k}_{i+3}$, $i = 1, 2, 3$. Note that any $\gamma \in \Gamma_1$ that acts nontrivially on one

Name	$(z_1, w_1; z_4, w_4)$
Travelling Rolls	$x(1,0;0,0)$
Travelling Squares	$x(1,0;1,0)$
Standing Rolls	$x(1,1;0,0)$
Standing Squares	$x(1,1;1,1)$
Alternating Rolls	$x(1,1;i,i)$

TABLE 1. The $\mathbb{C}$-axial solutions on the square lattice. Here $x \in \mathbb{C}$.

subsystem necessarily acts nontrivially on another. We add generators to Γ corresponding to certain translations and reflections of *individual* subsystems, so that each subsystem *can* be acted upon independently of the others. Each subsystem then has the local symmetry group $\mathbb{L} = T^2 + \mathbb{D}_4$ of functions with the periodicity of the square lattice. In addition we also have the global symmetry group $\mathbb{G} = S_3$, which permutes the subsystems. The spatial symmetry is thus the *wreath product* group $\mathbb{L} \wr \mathbb{G} = (T^2 + \mathbb{D}_4) \wr S_3$ and the group under which this new system is equivariant is $\Gamma_2 = \big[(T^2 + \mathbb{D}_4) \wr S_3\big] \times S^1$ (see [5, 7] for a definition of the wreath product).

For the wreath product $\mathbb{L} \wr \mathbb{G}$ we can determine the $\mathbb{C}$-axial subgroups of $(\mathbb{L} \wr \mathbb{G}) \times S^1$ [5]. This involves first finding the $\mathbb{C}$-axial subgroups of $\mathbb{L} \times S^1 = (T^2 + \mathbb{D}_4) \times S^1$, which is the Hopf bifurcation problem on the square lattice. This has been solved [10], and the solutions for the local subsystems are shown in Table 1. With these solutions we can use the method of [5] to construct the 21 $\mathbb{C}$-axial solutions for $\Gamma_2 \times S^1$, which are listed in Table 2. Notice that in any one solution all the subsystems either vanish or take one of the forms from the square lattice. The nonvanishing subsystems only differ by overall phase shifts.

It is easy to show that, for $\Gamma_1 \subset \Gamma_2$,

The fixed point subspace of any isotropy subgroup of Γ_1 is the fixed point subspace of *some* isotropy subgroup of Γ_2.

Thus every Γ_1-solution has a corresponding Γ_2-solution and we find all the $\mathbb{C}$-axial solutions on the BCC lattice.

A Γ_2-solution may correspond to no Γ_1-solution if a Γ_1-equivariant term breaks the invariance of the fixed point subspace. A Γ_2-solution may correspond to multiple Γ_1-solutions if the latter are conjugate by an element $\gamma \in \Gamma_2 \backslash \Gamma_1$. To find the Γ_1-solutions we take each Γ_2-solution $\mathbf{z}$, choose an arbitrary element $\gamma \in \Gamma_2 \backslash \Gamma_1$, and see which γ satisfy

$$\dim\Big[\mathrm{Fix}\big(\Sigma(\gamma \cdot \mathbf{z})\big)\Big] = 2.$$

The 21 solutions on the extended system yield 26 solutions on the BCC lattice, also shown in Table 2. Thus for example solution 5 of the extended system, namely $x(1,0;1,0\,|\,1,0;1,0\,|\,0,0;0,0)$, corresponds to no solution on the BCC lattice, while solution 2 corresponds to two.

	Γ_2-solution $\mathbf{z}$	Γ_1-solution $\mathbf{z}$
1	$x(1,0;0,0\,\|\,0,0;0,0\,\|\,0,0;0,0)$	$x(1,0;0,0\,\|\,0,0;0,0\,\|\,0,0;0,0)$
2	$x(1,0;0,0\,\|\,1,0;0,0\,\|\,0,0;0,0)$	$x(1,0;0,0\,\|\,1,0;0,0\,\|\,0,0;0,0)$
		$x(1,0;0,0\,\|\,0,1;0,0\,\|\,0,0;0,0)$
3	$x(1,0;0,0\,\|\,1,0;0,0\,\|\,1,0;0,0)$	$x(1,0;0,0\,\|\,1,0;0,0\,\|\,1,0;0,0)$
4	$x(1,0;1,0\,\|\,0,0;0,0\,\|\,0,0;0,0)$	$x(1,0;1,0\,\|\,0,0;0,0\,\|\,0,0;0,0)$
5	$x(1,0;1,0\,\|\,1,0;1,0\,\|\,0,0;0,0)$	
6	$x(1,0;1,0\,\|\,1,0;1,0\,\|\,1,0;1,0)$	
7	$x(1,1;0,0\,\|\,0,0;0,0\,\|\,0,0;0,0)$	$x(1,1;0,0\,\|\,0,0;0,0\,\|\,0,0;0,0)$
8	$x(1,1;0,0\,\|\,1,1;0,0\,\|\,0,0;0,0)$	$x(1,1;0,0\,\|\,1,1;0,0\,\|\,0,0;0,0)$
9	$x(1,1;0,0\,\|\,1,1;0,0\,\|\,1,1;0,0)$	$x(1,1;0,0\,\|\,1,1;0,0\,\|\,1,1;0,0)$
10	$x(1,1;0,0\,\|\,i,i;0,0\,\|\,0,0;0,0)$	$x(1,1;0,0\,\|\,i,i;0,0\,\|\,0,0;0,0)$
11	$x(1,1;0,0\,\|\,\zeta,\zeta;0,0\,\|\,\zeta^2,\zeta^2;0,0)$	$x(1,1;0,0\,\|\,\zeta,\zeta;0,0\,\|\,\zeta^2,\zeta^2;0,0)$
12	$x(1,1;1,1\,\|\,0,0;0,0\,\|\,0,0;0,0)$	$x(1,1;1,1\,\|\,0,0;0,0\,\|\,0,0;0,0)$
13	$x(1,1;1,1\,\|\,1,1;1,1\,\|\,0,0;0,0)$	$x(1,1;1,1\,\|\,1,1;1,1\,\|\,0,0;0,0)$
		$x(1,1;1,1\,\|\,1,1;-1,-1\,\|\,0,0;0,0)$
14	$x(1,1;1,1\,\|\,1,1;1,1\,\|\,1,1;1,1)$	$x(1,1;1,1\,\|\,1,1;1,1\,\|\,1,1;1,1)$
		$x(1,1;i,-i\,\|\,1,1;i,-i\,\|\,1,1;i,-i)$
15	$x(1,1;1,1\,\|\,i,i;i,i\,\|\,0,0;0,0)$	$x(1,1;1,1\,\|\,i,i;i,i\,\|\,0,0;0,0)$
		$x(1,1;1,1\,\|\,i,i;-i,-i\,\|\,0,0;0,0)$
16	$x(1,1;1,1\,\|\,\zeta,\zeta;\zeta,\zeta\,\|\,\zeta^2,\zeta^2;\zeta^2,\zeta^2)$	$x(1,1;1,1\,\|\,\zeta,\zeta;\zeta,\zeta\,\|\,\zeta^2,\zeta^2;\zeta^2,\zeta^2)$
		$x(1,1;i,-i\,\|\,\zeta,\zeta;i\zeta,-i\zeta\,\|\,\zeta^2,\zeta^2;i\zeta^2,-i\zeta^2)$
17	$x(1,1;i,i\,\|\,0,0;0,0\,\|\,0,0;0,0)$	$x(1,1;i,i\,\|\,0,0;0,0\,\|\,0,0;0,0)$
18	$x(1,1;i,i\,\|\,1,1;i,i\,\|\,0,0;0,0)$	$x(1,1;i,i\,\|\,1,1;i,i\,\|\,0,0;0,0)$
		$x(1,1;i,i\,\|\,1,1;-i,-i\,\|\,0,0;0,0)$
19	$x(1,1;i,i\,\|\,1,1;i,i\,\|\,1,1;i,i)$	$x(1,1;i,i\,\|\,1,1;i,i\,\|\,1,1;i,i)$
		$x(1,1;-1,1\,\|\,1,1;-1,1\,\|\,1,1;-1,1)$
20	$x(1,1;i,i\,\|\,\chi,\chi;\chi^3,\chi^3\,\|\,0,0;0,0)$	
21	$x(1,1;i,i\,\|\,\rho,\rho;i\rho,i\rho\,\|\,\rho^2,\rho^2;i\rho^2,i\rho^2)$	$x(1,1;i,i\,\|\,\rho,\rho;\rho,-\rho\,\|\,\rho^2,\rho^2;i\rho^2,i\rho^2)$
		$x(1,1;i,i\,\|\,\rho,\rho;-i\rho,-i\rho\,\|\,\rho^2,\rho^2;i\rho^2,i\rho^2)$

TABLE 2. The $\mathbb{C}$-axial solutions for the extended group Γ_2 (left) and the original group Γ_1 (right) with $\mathbf{z}$ written as in equation (∗). Here $\zeta = e^{i\pi/3}$, $\chi = e^{i\pi/4}$, $\rho = e^{i\pi/6}$ and $x \in \mathbb{C}$.

3. The Hopf bifurcation on the FCC lattice

The appropriate group extension for the FCC lattice is obtained by adding to Γ generators corresponding to translation and reflection of the single pair (z_1, w_1). Each subsystem (z_j, w_j) can then be translated and/or reflected independently. The resulting group is $O(2) \wr S_4$, the symmetry group of functions with the periodicity of the simple hypercubic (SH) lattice in four spatial dimensions. The $\mathbb{C}$-axial solutions on the SH and FCC lattices are then easily found [2].

We can also glean information about primary branches of *subaxial* solutions. The same group extension gives us the isotropy subgroups with four-dimensional fixed point subspaces [1]. Many of these have $\mathbb{D}_4$ symmetry, as for instance the subspaces

$$(x, x; y, y; 0, 0; 0, 0), \quad (x, x; y, y; ix, ix; iy, iy), \quad (x, y; x, y; x, y; x, y).$$

When restricted to these invariant subspaces the problem reduces to the Hopf bifurcation with square symmetry, for which primary quasiperiodic solutions are known to exist [11]. Additionally, the last of these three subspaces only has $\mathbb{D}_4$ symmetry by virtue of the normal form symmetry S^1. We thus expect this $\mathbb{D}_4$ to be weakly broken, which can lead to quite exotic dynamics [9].

References

[1] T. K. Callahan, in preparation

[2] T. K. Callahan, *Hopf bifurcations on the FCC lattice*, Proceedings of the Equa-diff99 Conference on Differential Equations, Berlin 1999, Vol. 1, 154–156 World Press Scientific, Singapore (2000)

[3] T. K. Callahan and E. Knobloch, *Bifurcations on the fcc lattice*, Phys. Rev. E **53**, 3559–3562 (1996)

[4] T. K. Callahan and E. Knobloch, *Symmetry-breaking bifurcations on cubic lattices*, Nonlinearity **10**, 1179–1216 (1997)

[5] A. P. S. Dias, *Hopf bifurcation for wreath products*, Nonlinearity, **11**, 2, 247–264 (1998)

[6] A. P. S. Dias and I. Stewart, *Hopf bifurcation on a simple cubic lattice*, Dynamics and Stability of Systems **14**, 3–55 (1999)

[7] B. Dionne, M. Golubitsky and I. Stewart, *Coupled cells with internal symmetry. I. Wreath products*, Nonlinearity, **9**, 2, 559–574 (1996)

[8] M. Golubitsky, I. Stewart, and D. G. Schaeffer, D. G., *Singularities and Groups in Bifurcation Theory vol. II.* Springer, Berlin (1988)

[9] J. Moehlis and E. Knobloch, *Forced symmetry breaking as a mechanism for bursting*, Phys. Rev. Lett. **80**, 5329–5332 (1998)

[10] M. Silber and E. Knobloch, *Hopf bifurcation on a square lattice*, Nonlinearity **4**, 1063–1107 (1991)

[11] J. W. Swift, *Hopf bifurcation with the symmetry of the square*, Nonlinearity **1**, 333–377 (1988)

T.K. Callahan
Department of Mathematics
Arizona State University
P.O. Box 871804
Tempe, AZ 85287-1804, USA
e-mail: timcall@math.la.asu.edu

Trends in Mathematics:
Bifurcations, Symmetry and Patterns, 129–134
© 2003 Birkhäuser Verlag Basel/Switzerland

Normal Forms of Dynamical Systems and Bifurcations

Giampaolo Cicogna

Abstract. We show the existence of a general class of bifurcating solutions to dynamical systems, by introducing their (Poincaré-Dulac) normal form, and imposing that the normalizing transformation is convergent. These bifurcating solutions include standard stationary and Hopf bifurcations, and multiple-periodic solutions as well.

1. The normalizing transformation.

Normal Form (NF) theory [1, 2, 4, 5] (see also [10], where many other references can be found) is an old subject in the study of Dynamical Systems (DS), having been introduced by Poincaré in his Thesis, and is still one of the most useful and used tools both in the qualitative and quantitative local analysis of DS. In this short paper, we will sketch an application of this method to bifurcation problems.

As well known, given a DS

$$\dot{x} \equiv \frac{dx}{dt} = f(x) ; \qquad x = x(t) \quad , \quad x \in \mathbf{R}^n , t \in \mathbf{R} \qquad (1.1)$$

where f is analytic in a neighourhood of a stationary point x_0 (i.e., a point such that $f(x_0) = 0$; we can choose $x_0 = 0$), the idea is to introduce a near-identity change of coordinates in order to eliminate the nonlinear terms in the given $f(x)$; the terms which cannot be eliminated, which are the "resonant terms", constitute the Poincaré-Dulac normal form, as we shall see more precisely in a moment.

Writing the DS (1.1) in the form

$$\dot{x} = f(x) = Ax + F(x) \qquad (1.2)$$

we shall always assume that the matrix $A = (Df)(0)$ is $\neq 0$ and semisimple, with eigenvalues λ_j; denoting by v the "new" coordinates, the NF of (1.1) will be written

$$\dot{v} = \widehat{f}(v) = Av + \widehat{F}(v) \qquad (1.3)$$

(the notation $\widehat{}$ will be always reserved to NF).

The coordinate transformation $x \to v$ is usually performed by means of recursive techniques, but in general this normalizing transformation (NT) is actually purely formal, and only very special conditions can ensure its convergence and the (local) analyticity of the NF [1, 2, 4, 5].

For our present applications, we will resort in particular to two very general conditions, given by Bruno, which ensure the convergence of the NT, and are called Condition A and Condition ω (see [4, 5] for details). These conditions read:

Condition A: A normal form $\widehat{f}$ is said to satisfy Condition A if $\widehat{f}$ has the form

$$\widehat{f}(v) = Av + \alpha(v)Av \qquad (1.4)$$

where $\alpha(v)$ is some scalar-valued power series $\big($with $\alpha(0) = 0\big)$.

Condition ω: Let $\omega_k = \min |(Q, \Lambda) - \lambda_j|$, $\forall j = 1, \ldots, n$ and n−tuples of integers $q_i \geq 0$ such that $1 < \sum_{i=1}^{n} q_i < 2^k$ and $(Q, \Lambda) = \sum_i q_i \lambda_i \neq \lambda_j$: then Condition ω is satisfied if

$$\sum_{k=1}^{\infty} 2^{-k} \ln \left(\omega_k^{-1} \right) < \infty \qquad (1.5)$$

Let us point out that we have stated here Condition A in its simplest (and quite restrictive) form, which holds, in particular, when the eigenvalues are either all real or all pure imaginary (which is just the case we will deal with). Whereas Condition A is clearly a quite strong condition on normal forms, Condition ω is a very weak arithmetic condition on the eigenvalues of the matrix A, generalizing Siegel-type conditions on the appearance of small divisors, and which is satisfied by almost all (in the Lebesgue sense) n−tuples of eigenvalues. We will assume for the sake of simplicity that it is always satisfied; let us also notice that, in some cases, Condition ω can actually be relaxed [17].

We can then state [4, 5]:

Theorem 1.1. (Bruno) *If $A = (Df)(0)$ satisfies Condition ω, and if f can be transformed, via a series of coordinate transformations, to a $\widehat{f}$ which satisfies Condition A, then there is a NT for f which is convergent in some open neighbourhood of $x_0 = 0$.*

The presence of "resonant terms", i.e., of the terms which constitute the nonlinear part $\widehat{F}(v)$ and which cannot be eliminated by the NT, is related to the existence of some "resonance" between the eigenvalues λ_i of the matrix A, i.e., to the existence of some non-negative integers m_i such that, for some index j:

$$\sum_{i=1}^{n} m_i \lambda_i = \lambda_j , \quad 1 \leq j \leq n \quad ; \quad \sum_{i=1}^{n} m_i \geq 2 . \qquad (1.6)$$

The form of the resonant terms is then given by the following result [1, 10, 11, 16]:

Lemma 1.2. *Given the matrix A, the most general NF has the form*

$$\widehat{F}(v) = \sum_j \mu_j(v) M_j v \quad \text{with} \quad \mu_j(v) \in \mathcal{I}_A \quad \text{and} \quad [M_j, A] = 0 \qquad (1.7)$$

where the sum is extended to a set of linearly independent matrices M_j commuting with A (the set of these matrices clearly includes A), and $\mathcal{I}_A$ denotes the set of the meromorphic constants of motion (or first integrals) of the linear problem $\dot{v} = Av$.

2. A "reduction lemma" for NF.

In this and the next sections, we will apply the previously sketched ideas to show the existence of bifurcating solutions to dynamical systems, depending on one or more real "control" parameters $\eta \in \mathbf{R}^p$. Precisely, we are now considering DS of the form

$$\dot{x} = f(x, \eta) \equiv A(\eta)x + F(x, \eta) \qquad (2.1)$$

where $f(x, \eta)$ is analytic in a neighbourhood of $x_0 = 0$, $\eta_0 = 0$, with $f(0, \eta) = 0$, and assume that, for some "critical" value $\eta = \eta_0 = 0$ of the parameters, the matrix

$$A_0 = A(0) \qquad (2.2)$$

is semisimple and its eigenvalues λ_i satisfy a resonance relation (1.6). Before giving our main result (next section), let us point out an useful application of the above methods. Indeed, the notions of normal form and of resonance, thanks in particular to Lemma 1.2, may directly lead to a "reduction lemma" for NF, allowing to a reduction of the original problem to a lower dimensional case. We have precisely (see also [3]):

Lemma 2.1. *Consider a n−dimensional DS ($n > 2$), and assume that for $\eta_0 = 0$ there are $r < n$ resonant eigenvalues, say $\lambda_1, \ldots, \lambda_r$, such that no resonance relation of the following form*

$$\sum_{h=1}^{r} m_h \lambda_h = \lambda_k \qquad k = r+1, \ldots, n \qquad (2.3)$$

exists. Then, the NF of this DS "splits" the variables $v_1, \ldots, v_r$ from the remaining $n - r$ variables in such a way that this NF admits a solution where

$$v_k(t) \equiv 0 \qquad k = r+1, \ldots, n \qquad (2.4)$$

Then, the DS – once in NF – can be reduced to a r−dimensional problem, and there is an invariant manifold under the dynamical flow of (2.1), which corresponds to the hyperplane (2.4) of the normal form.

Although this result is based on completely different arguments, this situation looks quite similar to the case of the "equivariant bifurcation lemma". E.g., if $\lambda_1 = 0$, $r = 1$, or resp. $\lambda_1 = -\lambda_2 = i$, $r = 2$, and condition (2.3) is satisfied, the NF turns out to be resp. 1−dimensional and 2−dimensional, in some similarity to the situation met in the equivariant lemma (resp. in the stationary bifurcation [6, 15], and in the Hopf bifurcation [7, 12]; see [13] for a comprehensive discussion).

Once a solution $v_h(t)$, $h = 1, \ldots, r$ of this reduced r–dimensional problem has been found, then the original DS admits a solution in which the $n - r$ components $x_k(t)$, $k = r + 1, \ldots, n$, are "higher-order terms" with respect to the first r components $x_h(t)$, showing here some analogy with the Lyapunov-Schmidt procedure.

3. An application of convergent NF: the "resonant bifurcation".

With the above positions, we can now give a theorem [8] ensuring the existence – under suitable hypotheses – of a general class of bifurcating solutions in correspondence to resonant point $\eta_0 = 0$. The main idea is to transform the given DS into NF and to impose that the NT is convergent, using the convergence conditions of Theorem 1.1; or, more precisely, to take advantage from the presence of the parameters η in order to impose that the general NF as given by (1.6) takes the special form (1.4), as required by Condition A. Thanks to the remarks in the above section, it is not restrictive to assume the resonance involves all the eigenvalues $\lambda_1, \ldots, \lambda_n$ of the matrix A_0.

Theorem 3.1. *Consider the DS* (2.1) *and assume that for the value* $\eta_0 = 0$ *the eigenvalues* λ_i *of* A_0 *are distinct, real or purely imaginary, and satisfy a resonance relation* (1.5). *Assume also that* $p = n - 1$, *and finally that putting*

$$a_k^{(i)} = \left. \frac{\partial A_{ii}(\eta)}{\partial \eta_k} \right|_{\eta=0} \qquad (i = 1, \ldots, n \; ; \; k = 1, \ldots, n - 1) \qquad (3.1)$$

the following $n \times n$ *matrix* D *is not singular, i.e., that:*

$$\det D \equiv \det \begin{pmatrix} \lambda_1 & a_1^{(1)} & a_2^{(1)} & \cdots & a_{n-1}^{(1)} \\ \lambda_2 & a_1^{(2)} & \cdots & \\ \cdots & & & \\ \lambda_n & a_1^{(n)} & \cdots & a_{n-1}^{(n)} \end{pmatrix} \neq 0 \qquad (3.2)$$

Then, there is, in a neighbourhood of $x_0 = 0$, $\eta_0 = 0$, $t = 0$, *a bifurcating solution of the form*

$$x_i(t) = \Big[\exp \big((1 + \beta(\eta)) A_0 t) \big) \Big] x_{0i}(\eta) + \text{h.o.t.} \qquad i = 1, \ldots, n$$

where $\beta(\eta)$ *is some function of the* η's *such that* $\beta(\eta) \to 0$ *for* $\eta \to 0$, *and h.o.t. stands for higher order terms vanishing for* $\eta \to 0$.

Notice that standard stationary bifurcation, Hopf bifurcation, and multiple-periodic bifurcating solutions as well, are particular cases of the bifurcations obtained in this way. For instance, if $n = 2$ and with imaginary eigenvalues, it is easy to see that condition (3.2) coincides with the well known "transversality condition" $d \operatorname{Re}\lambda(\eta)/d\eta|_{\eta=0} \neq 0$ ensuring standard Hopf bifurcation. A nontrivial example in dimension $n > 2$, and corresponding to the case of coupled oscillators with multiple frequencies, is covered by this Corollary, which immediately follows from the previous Theorem.

Corollary 3.2. *With the same notations as before, let $n = 4$ and $\lambda_1 = -\lambda_2 = i\omega_0$, $\lambda_3 = -\lambda_4 = mi\omega_0$ (with $m = 2, 3, \ldots$): with $\eta \equiv (\eta_1, \eta_2, \eta_3) \in \mathbf{R}^3$, let, after complexification of the space, $A^C(\eta)$ be the matrix $A(\eta)$ written in the basis in which $A^C(0)$ is diagonal. Putting $a_k^{(i)} = \left. \dfrac{\partial A_{ii}^C(\eta)}{\partial \eta_k} \right|_{\eta=0}$ $(i = 1, \ldots, 4 ; \; k = 1, 2, 3)$, assume that*

$$\det D = \det \begin{pmatrix} a_1^{(1)} & a_2^{(1)} & a_3^{(1)} & 1 \\ a_1^{(2)} & \cdots & & -1 \\ \cdots & \cdots & & m \\ a_1^{(4)} & \cdots & & -m \end{pmatrix} \neq 0 \tag{3.2'}$$

then there is a multiple-periodic bifurcating solution preserving the frequency resonance $1 : m$.

Just to give briefly an example of the case covered by this Corollary, assume that

$$A(\eta) = \begin{pmatrix} \eta_1 + \eta_3 & -(1 + \eta_2) & \eta_1 & 0 \\ 1 + \eta_2 & \eta_1 - \eta_3 & 0 & \eta_1 \\ \eta_2 & 0 & \eta_3 & 2 \\ 0 & -\eta_2 & 2 & \eta_3 \end{pmatrix}$$

which satisfies $(3.2')$, with $m = 2$ and $\omega_0 = 1$; the constants of the motion entering in the NF at the resonance $\eta = 0$ (cf. Lemma 1.2) are

$$\mu_1 = r_a^2 = v_1^2 + v_2^2 \; ; \quad \mu_2 = r_b^2 = v_3^2 + v_4^2 \; ; \quad \mu_3 = 2r_a^2 r_b \cos 2\varphi$$

where φ is the time phase-shift between the two first two components v_1, v_2 and the remaining v_3, v_4. It is easily seen (see [8] for details) that the leading terms of the bifurcating solution can be written

$$x_1 + ix_2 = r_a \exp(i\omega t) \; ; \quad x_3 + ix_4 = r_b \exp(2i\omega(t + \varphi))$$

where $\omega = 1 + \beta$, and β, r_a, r_b, φ are connected to the parameters η_i by some relationships which depend on the explicit form of the nonlinear part of $F(x, \eta)$ of the DS (2.1).

The above results can be suitably extended [8, 9] to the case of multiple eigenvalues of the matrix A_0, in the presence of some general symmetry property of the problem (including the case of local or non linear Lie-point symmetries [10, 14]).

An example of this situation, given by coupled oscillators with degenerate frequencies and in the presence of a rotation symmetry, is described in [8].

References

[1] V.I. Arnold, *Geometrical methods in the theory of differential equations*, (1988) Springer, Berlin

[2] V.I. Arnold and Yu.S. Il'yashenko, *Ordinary differential equations*; in: Encyclopaedia of Mathematical Sciences vol. 1 – Dynamical Systems I, (D.V. Anosov and V.I. Arnold eds.), 1–148 (1988) Springer, Berlin

[3] Yu.N. Bibikov, *Local theory of nonlinear analytic ordinary differential equations*, (1979) Springer, Berlin

[4] A.D. Bruno, *Analytical form of differential equations*, Trans. Moscow Math. Soc., **25** (1971), 131–288; and **26** (1972), 199–239

[5] A.D. Bruno, *Local methods in the theory of differential equations*, (1989) Springer, Berlin

[6] G. Cicogna, *Symmetry breakdown from bifurcation*, Lett. Nuovo Cim. **31** (1981), 600–602

[7] G. Cicogna, *Two ways for Hopf bifurcation with symmetry*, Journ. Phys. A **19** (1986), L369–L371

[8] G. Cicogna, *Resonant bifurcations*, J. Math. Anal. Appl., **241** (2000), 157–180

[9] G. Cicogna, *Convergent normal forms, symmetries, and applications to bifurcations*, Intern. J. Dyn. Cont. Discrete and Impulsive Systems, to appear

[10] G. Cicogna and G. Gaeta, *Symmetry and Perturbation Theory in Nonlinear Dynamics*, (1999) Springer, Berlin

[11] C. Elphick, E.Tirapegui, M.E. Brachet, P. Coullet, and G. Iooss, *A simple global characterization for normal forms of singular vector fields*, Physica D, **29** (1987), 95–127

[12] M. Golubitsky and I. Stewart, *Hopf bifurcation in the presence of symmetry*, Arch. Rat. Mech. Anal. **87** (1985), 107–165

[13] M. Golubitsky, D. Schaeffer and I. Stewart, *Singularities and groups in bifurcation theory – vol. II* (1988), Springer, Berlin

[14] P.J. Olver, *Applications of Lie groups to differential equations* (1986, and second edition 1993) Springer, Berlin

[15] A. Vanderbauwhede, *Local bifurcation and symmetry*, (1982) Pitman, Boston

[16] S. Walcher, *Differential equations in normal form*, Math. Ann., **291** (1991), 293–314

[17] S. Walcher, *On convergent normal form transformations in presence of symmetries*, J. Math. Anal. Appl., **244** (2000), 17–26

Giampaolo Cicogna
Dipartimento di Fisica
Università di Pisa
Via Buonarroti 2, Ed. B
I-56127 Pisa, Italy
e-mail: `cicogna@df.unipi.it`

Trends in Mathematics:
Bifurcations, Symmetry and Patterns, 135–140
© 2003 Birkhäuser Verlag Basel/Switzerland

One-dimensional Pattern Formation in Systems with a Conserved Quantity

S.M. Cox and P.C. Matthews

Abstract. Regular one-dimensional patterns in systems with a reflection symmetry and a conserved quantity may be unstable to an instability leading to strong spatial modulation of the pattern. For certain parameter values, *all* regular patterns may be unstable at onset; simulations then indicate the existence of stable strongly modulated patterns. Analysis of the instability has hitherto assumed that the linear growth rate of disturbances is $O(k^2)$ as the wavenumber $k \to 0$. However, the instability is shown here to be present even when there is slight damping of the modes with $k \to 0$, corresponding to a slight breaking of the conservation law.

1. Introduction and derivation of amplitude equations

We consider one-dimensional pattern formation in systems with a conserved quantity. The presence of a conservation law leads to the existence of a slowly-evolving large-scale mode, whose behaviour may significantly affect pattern formation in the system. The effects of the large-scale mode persist, even when this mode is slightly damped, as we demonstrate in this paper.

We suppose that the system under consideration has a uniform state which becomes unstable, as a bifurcation parameter r passes through the bifurcation value r_c, to a pattern with wavenumber $k_c > 0$. To further simplify matters, we assume the existence of a reflection symmetry in the spatial variable, x, and that the onset of pattern formation takes place through a stationary bifurcation. There are then two ways in which the conserved quantity may behave under the reflection symmetry: it may either change sign (see the paper in this volume by Matthews and Cox, and the companion paper [8]) or not (the case considered here, and in the companion paper [7]).

We consider a partial differential equation (PDE) for a conserved quantity $w(x,t)$, with a uniform solution, which may without loss of generality be taken to be the trivial solution $w = 0$. Suppose that the system has the reflection and translation symmetries $x \mapsto -x$, $w \mapsto w$ and $x \mapsto x + x_0$, $w \mapsto w$. Then all terms in the linearised expression for $\partial w/\partial t$ contain a (nonzero) even number of

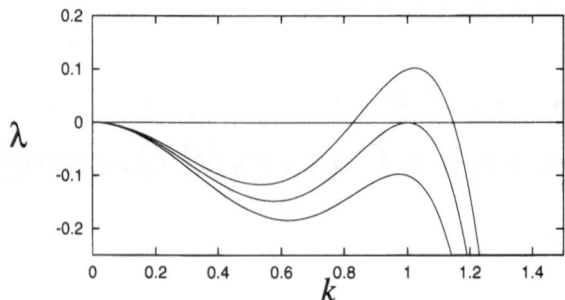

FIGURE 1. Linear growth rate λ in (1), plotted as a function of wavenumber k, for $r = -0.1$ (bottom curve), 0 and 0.1 (top curve). This is typically the qualitative behaviour of the linear growth rate for systems such as are envisaged here.

x-derivatives. An appropriate partial differential equation is

$$\frac{\partial w}{\partial t} = -\frac{\partial^2}{\partial x^2}\left[rw - \left(1 + \frac{\partial^2}{\partial x^2}\right)^2 w - sw^2 - w^3\right], \tag{1}$$

where r and s are parameters. For this equation, the growth rate of small disturbances with wavenumber k is $\lambda = k^2(r - (1 - k^2)^2)$. Thus for small positive values of the bifurcation parameter r, a narrow band of wavenumbers near $k = k_c = 1$ exhibits slow linear growth (see figure 1), while modes with small wavenumber also evolve slowly. Although one might imagine that the large-scale modes have little influence on the pattern because they decay according to linear theory, this turns out to be far from the case. (In the next section we shall show that the influence of the large-scale modes persists even when λ is strictly negative for the modes with $k = 0$.)

Whether we choose to focus on (1) or on a different PDE, we assume with no loss of generality that lengths are scaled so that $k_c = 1$. Then near the onset of pattern formation w takes the form

$$w(x,t) \sim \tilde{A}(X,T)e^{ix} + \tilde{A}^*(X,T)e^{-ix} + \tilde{B}(X,T),$$

where X and T are rescaled forms of x and t, and $\tilde{A}$ and $\tilde{B}$ are small amplitudes (the asterisk denotes complex conjugation). By considering the symmetries of the PDE, and assuming direct quadratic coupling between the pattern and the large-scale mode in the PDE, we find that the amplitude equations for A and B take the form

$$A_T = A + A_{XX} - A|A|^2 - AB, \tag{2}$$

$$B_T = \sigma B_{XX} + \mu(|A|^2)_{XX}, \tag{3}$$

where $r = \epsilon^2$, $\tilde{A} = \epsilon A$, $\tilde{B} = \epsilon^2 B$, $X = \epsilon x$ and $T = \epsilon^2 t$. Although no details are given here, we note that the reduction of a PDE (or system of PDEs) to the amplitude equations (2) and (3) can be carried out in an asymptotically consistent fashion. Furthermore, in deriving (2) and (3), we have assumed that the coefficient of $A|A|^2$ can be scaled to -1, i.e., that in the absence of B the bifurcation of the pattern mode is supercritical. All coefficients in (2) have been scaled to unity, leaving two coefficients in (3) that cannot be removed by rescaling. In a particular application, σ and μ may be computed in terms of the system parameters, through a weakly nonlinear expansion of the solution. For the purposes of the present discussion, however, it suffices to note that slow damping of the large-scale modes implies that $\sigma > 0$, but μ may take either sign. The equations (2) and (3) have previously been derived to describe the coupled dynamics of sand waves and sand banks [6], and to describe the *secondary* stability of cellular patterns [1].

The amplitude equations (2) and (3) hold near the stationary onset of pattern formation in any system in one space dimension with translational and reflectional symmetries, and with a conservation law, provided that there is an appropriate quadratic coupling.

2. Solutions to the amplitude equations

Roll solutions to the governing PDE with wavenumber $k_c + \epsilon q$ correspond to solutions to the amplitude equations (2) and (3) of the form $A = (1 - q^2)^{1/2}e^{iqX}$, $B = 0$. It is straightforward to investigate the stability of these rolls to disturbances with wavenumber $k_c + \epsilon(q \pm l)$ [7]. Both Hopf and stationary bifurcations may take place, but we focus on the latter as we have observed them in numerical simulations. The most dangerous modes are those with $l \to 0$, in which case rolls are unstable if $\mu(1 - q^2) > \sigma(1 - 3q^2)$. Hence *all* rolls are unstable if $\mu > \sigma$.

In the limit $|\sigma/\mu| \gg 1$, where the large-scale mode is strongly damped, the Eckhaus criterion $q^2 > \frac{1}{3}$ for instability [4] is recovered; the band of stable rolls is narrowed or broadened when μ is positive or negative, respectively. For general values of the wavenumber of the rolls, the instability has characteristics of both amplitude and phase instabilities, and for $q \neq 0$ is generically subcritical, at least in a sufficiently long domain [9]. However, for rolls with exactly critical wavenumber, the bifurcation is predicted to be supercritical. Indeed, in numerical simulations we have found a smooth transition from uniform to modulated rolls as μ/σ is increased through unity, indicative of a supercritical bifurcation.

Our analysis so far has assumed that the linear decay rate vanishes for the modes of largest wavelength. In practice, however, this property may hold only approximately (if $w(x, t)$ is only approximately conserved, for instance). In this case, the amplitude equations (2) and (3) become

$$A_T = A + A_{XX} - A|A|^2 - AB \tag{4}$$
$$B_T = -\gamma B + \sigma B_{XX} + \mu(|A|^2)_{XX}, \tag{5}$$

where $\gamma > 0$. Such equations are appropriate, for example, for thermal convection between boundaries at which the heat flux is given by Newton's law of cooling

(a) $\mu' < 0$
(b) $0 < \mu' < 3$ $\frac{4}{3}\mu' < \gamma'$
(c) $3 < \mu'$ $2(\mu' - 1) < \gamma'$
(d) $0 < \mu' < 1$ $\gamma' < \frac{4}{3}\mu'$ $q^2 < q_c^2$
(e) $1 < \mu' < 3$ $2(\mu' - 1) < \gamma' < \frac{4}{3}\mu'$ $q^2 < q_c^2$
(f) $3 < \mu'$ $\frac{4}{3}\mu' < \gamma' < 2(\mu' - 1)$ $q_c^2 < q^2 < \frac{1}{3}$

TABLE 1. *Sufficient* conditions for roll solutions of (4) and (5) to be *stable* to monotonic disturbances (where condition (8) is broken). In each of cases (d), (e) and (f), $q_c^2 = \frac{1}{2}(2(1 - \mu') + \gamma')/(3 - \mu')$.

(heat flux proportional to the difference between the boundary temperature and some reference temperature). The criterion for monotonic instability is then

$$\sigma l^4 + (\gamma - 2\mu a_0^2 + 2\sigma a_0^2 - 4\sigma q^2)l^2 - 2\gamma(2q^2 - a_0^2) < 0, \tag{6}$$

where $a_0 = (1 - q^2)^{1/2}$. Rolls are thus stable to disturbances with large l, while the Eckhaus criterion for instability [4] is recovered in the limit $l \to 0$. This criterion is also recovered in the limit $\gamma \to \infty$; this is not surprising since it reflects enslavement of the mean mode to the pattern mode, according to $B \sim \mu(|A|^2)_{XX}/\gamma$, which is the usual situation in the absence of a conservation law.

Since rolls with $\frac{1}{3} < q^2 < 1$ are Eckhaus unstable, it is of interest to investigate the stability of rolls with $q^2 < \frac{1}{3}$, which are Eckhaus stable. Although such rolls are stable in the limit $l \to 0$, they may be unstable to finite-l disturbances. For general values of l, there is monotonic instability if the two conditions

$$0 > 8\gamma'(1 - 3q^2) - [\gamma' - 2\mu'(1 - q^2) + 2(1 - 3q^2)]^2, \tag{7}$$
$$0 > \gamma' - 2\mu'(1 - q^2) + 2(1 - 3q^2) \tag{8}$$

are simultaneously satisfied, where $\gamma' = \gamma/\sigma$ and $\mu' = \mu/\sigma$. Condition (8) is broken in the cases listed in table 1, which thus provide sufficient conditions for the rolls to be stable to monotonic disturbances with any value of l. The inequalities (7) and (8) are most easily analysed when $q = 0$: this mode is unstable whenever $\mu' > 1 + \gamma'/2 + (2\gamma')^{1/2}$ (consistent with the condition $\mu' > 1$ in the absence of damping, derived above).

3. Application to rotating convection

The instability described above can be found [2] in two-dimensional thermal convection in a horizontal fluid layer rotating about a vertical axis, if the top and bottom boundaries of the layer are stress-free. Then the component of fluid momentum along the roll axes is the conserved quantity. A weakly nonlinear analysis leading to equations of the form (2) and (3) allows us to map the location of the instability in Taylor number–Prandtl number space. The instability sets in as the

FIGURE 2. Contours of perturbation temperature in rotating convection at Taylor number $4\pi^4$ and (supercritical) Rayleigh number 1286. The Prandtl number is, from the top, 1.0, 0.95, 0.9, 0.8, 0.7. The horizontal boundaries are stress-free and isothermal.

Prandtl number decreases through some critical value, dependent on the Taylor number; at Taylor number $4\pi^4$ the instability is most dangerous, and arises as the Prandtl number decreases through unity. The corresponding localisation of the convection rolls can be seen in the numerical results shown in figure 2, for a computational domain chosen to fit ten pairs of rolls with critical wavenumber. For $Pr = 1$, the regular rolls are marginally stable to the instability described above, and the computed rolls are unmodulated (top row); for smaller values of Pr, the rolls are unstable, and the corresponding long-time steady state is shown in the bottom four rows of figure 2, for a variety of Prandtl numbers.

4. Discussion

Two questions asked after the presentation of this material at the Conference concerned the extension of these ideas to oscillatory onset, and to patterns in two space dimensions.

The extension to the case of oscillatory onset presents some analytical subtleties, since there will be leftward and rightward travelling waves, each having an envelope that moves with its own group velocity [5], necessitating consideration of two distinct frames of reference; in fact a third frame (the stationary one associated with the large-scale mode) is also needed.

The amplitude-equation approach as adopted here may also be applied to problems in higher dimensions [3]. Some localised solutions in higher dimension have been found experimentally by VanHook *et al.* [10, 11]; these support the theoretical exploration of the higher-dimensional case.

References

[1] P. Coullet and G. Iooss, *Instabilities of one-dimensional cellular patterns*, Phys. Rev. Lett., **64** (1990), 866–869.

[2] S. M. Cox and P. C. Matthews, *New instabilities in two-dimensional rotating convection and magnetoconvection*, Physica D, **149** (2001), 210–229.

[3] S. M. Cox and P. C. Matthews, *Instability and localisation of patterns due to a conserved quantity*, (2001) preprint.

[4] W. Eckhaus, *Studies in Non-linear Stability Theory*, 1965, Springer-Verlag, New York.

[5] E. Knobloch and J. De Luca, *Amplitude equations for travelling wave convection*, Nonlinearity, **3** (1990), 975–980.

[6] N. L. Komarova and A. C. Newell, *Nonlinear dynamics of sand banks and sand waves*, J. Fluid Mech., **415** (2000), 285–321.

[7] P. C. Matthews and S. M. Cox, *Pattern formation with a conservation law*, Nonlinearity, **13** (2000), 1293–1320.

[8] P. C. Matthews and S. M. Cox, *One-dimensional pattern formation with Galilean invariance near a stationary bifurcation*, Phys. Rev. E, **62** (2000), R1473–R1476.

[9] M. R. E. Proctor, *Finite amplitude behaviour of the Matthews–Cox instability*, (2001) preprint.

[10] S. J. VanHook, M. F. Schatz, W. D. McCormick, J. B. Swift and H. L. Swinney, *Long-wavelength instability in surface-tension-driven Bénard convection*, Phys. Rev. Lett., **75** (1995), 4397–4400.

[11] S. J. VanHook, M. F. Schatz, W. D. McCormick, J. B. Swift and H. L. Swinney, *Long-wavelength surface-tension-driven Bénard convection: experiment and theory*, J. Fluid Mech., **345** (1997), 45–78.

S.M. Cox and P.C. Matthews
School of Mathematical Sciences
University of Nottingham
University Park
Nottingham NG7 2RD, UK
e-mail: stephen.cox@nottingham.ac.uk
e-mail: paul.matthews@nottingham.ac.uk

Trends in Mathematics:
Bifurcations, Symmetry and Patterns, 141–150
© 2003 Birkhäuser Verlag Basel/Switzerland

Invariance and Symmetry in a Year-class Model

O. Diekmann and S.A. van Gils

Abstract. In this note we reveal some of the special structure of a year-class model. We formulate a certain parameter symmetry and compute the characteristic equation at the unique nontrivial equilibrium. In the case of equal sensitivity we derive phase-amplitude equations and show the existence of an invariant manifold.

1. Introduction

Population models yield non-generic dynamical systems, since necessarily any subspace defined by the absence of a species is invariant. The aim of this paper is to uncover the enhanced non-genericity in discrete time models for populations consisting of individuals that reproduce only once in their life (such species are called semelparous or, in the case of insects, univoltine. Examples range from annual and biennial plants to salmon to cicadas).

If reproduction is restricted to a small time window in the year and life span has a fixed length of, say, k years, a population splits into year-classes according to the year of birth (mod k) or, equivalently, the year of reproduction (mod k). As a year-class is reproductively isolated from the other year-classes, it forms a population by itself.

Yet year-classes are likely to interact, for instance by competition for food. It then may happen that a year-class is driven to extinction. Bulmer [Bul77] calls an insect *periodical* if it consists of a single year-class, i.e., if all but one year-classes are missing. Famous examples are the cicada species with 13 and 17 year life cycles (see, e.g., [Beh00] and the references given there).

Mathematically the phenomenon of possibly missing year-classes shows up as *invariance* of coordinate axes and (hyper)planes for the "full life cycle" map of looking k years ahead. For each invariant subspace we can investigate the dynamics within it, as well as the *external stability*, by which we mean the attraction or repulsion in the transverse direction (or, in more biological terms, the decline or growth of a missing year-class which is introduced in small numbers).

We shall describe interaction (i.e., density dependence) as feedback via a one dimensional environmental quantity. Moreover, we take the various feedback relations to be of the same functional form, i.e., they differ only by a scaling parameter. See the next section for the details. Some of the features we find are due to this form of density dependence, rather than semelparity per se.

In this paper we shall reveal some of the special structure of the model. We formulate a certain parameter symmetry, i.e., an equivariance under transformations of both the state variables *and* the parameters. We will also compute the characteristic equation at the unique equilibrium. The characteristic equation is the starting point when trying to determine the stability of the equilibrium. In this preliminary presentation we do not yet touch upon important questions like: Can one year-class tune the environmental conditions such that the other year-classes are driven to extinction when rare? Or can missing year-classes invade successfully? Do we get coexistence or competitive exclusion? A rather complete answer to these questions in the case of *biennials*, i.e., $k = 2$, is given in [DDvG].

2. The model

In this paper we consider the nonlinear Leslie matrix iteration

$$N(t+1) = L(I(t)) N(t), \tag{2.1}$$

where t is an integer(!) and

$$N(t) = \begin{pmatrix} N_0(t) \\ \vdots \\ N_{k-1}(t) \end{pmatrix}, \tag{2.2}$$

so the components of $N(t)$ measure the size of the k different age-classes at time t, and

$$L(I) = R_0^{\frac{1}{k}} \begin{pmatrix} 0 & & & & e^{-g_{k-1} I} \\ e^{-g_0 I} & & & & \\ & e^{-g_1 I} & & & \\ & & \ddots & & \\ & & & e^{-g_{k-2} I} & 0 \end{pmatrix} \tag{2.3}$$

$$I = c \cdot N = c_0 N_0 + \ldots + c_{k-1} N_{k-1}, \tag{2.4}$$

with normalization of the parameters

$$\sum_{i=0}^{k-1} g_i = \sum_{i=0}^{k-1} c_i = 1. \tag{2.5}$$

The formulation above is obtained by rescaling of the equations

$$
\begin{cases}
\tilde{N}_0(t+1) = f s_{k-1} e^{-\tilde{g}_{k-1} I(t)}\, \tilde{N}_{k-1}(t) \\[2mm]
\tilde{N}_1(t+1) = \ \ s_0 \ \ e^{-\tilde{g}_0\, I(t)}\, \tilde{N}_0(t) \\[2mm]
\qquad\qquad \vdots \\[6mm]
\tilde{N}_{k-1}(t+1) = \ \ s_{k-2} e^{-\tilde{g}_{k-2}\, I(t)}\, \tilde{N}_{k-2}(t)
\end{cases}
\tag{2.6}
$$

where $s_0 \ldots s_{k-1}$ are survival probabilities under "ideal" conditions, f is the expected number of offspring of a reproducing individual (again under "ideal" conditions) and $\exp(-\tilde{g}_i I(t))$ is the reduction of the survival probability in the $i+1$-th year due to crowding.

We do not specify what exactly are the adverse effects of crowding, but we simply postulate that their effect is described well by the factors $\exp(-\tilde{g}_i I(t))$, where

$$
I(t) = \tilde{c}_0 \tilde{N}_0(t) + \cdots + \tilde{c}_{k-1} \tilde{N}_{k-1}(t),
\tag{2.7}
$$

i.e., the coefficients $\tilde{c}_i$ measure the contribution of an i-year old individual to the reduction of survival.

The rescaling is given by

$$
\tilde{N}_i = \frac{\alpha_i N_i}{(\tilde{g}_0 + \cdots + \tilde{g}_{k-1})(\tilde{c}_0 \alpha_0 + \cdots + \tilde{c}_{k-1} \alpha_{k-1})},
\tag{2.8}
$$

where

$$
\alpha_0 = f^{\frac{k-1}{k}} s_{k-1}^{\frac{k-1}{k}}
$$

$$
\alpha_i = \frac{s_0 \cdot \ldots \cdot s_{i-1} \alpha_0^{i+1}}{f^{2i} s_{k-1}^{i}}, \quad i = 1 \ldots k-1
$$

$$
g_i = \frac{\tilde{g}_i}{\tilde{g}_0 + \ldots + \tilde{g}_{k-1}}, \quad i = 0 \ldots k-1
\tag{2.9}
$$

$$
c_i = \frac{\tilde{c}_i \alpha_i}{\tilde{c}_0 \alpha_0 + \ldots + \tilde{c}_{k-1} \alpha_{k-1}}, \quad i = 0 \ldots k-1
$$

$$
R_0 = f \prod_{i=0}^{k-1} s_i.
$$

3. Parameter symmetry

We will use the following notations. The shift on $\mathbb{R}^k$ is denoted by S:

$$S\,x = S\begin{pmatrix} x_0 \\ \vdots \\ x_{k-1} \end{pmatrix} = \begin{pmatrix} x_{k-1} \\ x_0 \\ \vdots \\ x_{k-2} \end{pmatrix}. \tag{3.10}$$

For $a \in \mathbb{R}^k$, $\mu \in \mathbb{R}$ we let

$$D(a,\mu) = \mathrm{diag}\left(e^{-a_0\mu},\dots,e^{-a_{k-1}\mu}\right). \tag{3.11}$$

Using this notation we can rewrite (2.1) in the form

$$N(t+1) = R_0^{\frac{1}{k}}\, D(Sg, c \cdot N(t))\, SN(t). \tag{3.12}$$

In this form it is easy to demonstrate the next

Proposition 3.1. *The system* (2.1) *is equivariant under the transformation* $N \mapsto SN$, $g \mapsto Sg$, $c \mapsto Sc$.

Proof. If $\mathrm{diag}(a_0,\dots,a_{k-1})$ is a diagonal matrix with the components of $a \in \mathbb{R}^k$ on the diagonal, then under similarity transformation with S the elements on the diagonal are shifted, i.e., $S\,\mathrm{diag}(a)\,S^{-1} = \mathrm{diag}(Sa)$. Hence, from (3.12) we infer

$$\begin{aligned} S\,N(t+1) &= R_0^{\frac{1}{k}}\, S\, D(Sg, c \cdot N(t))\, S^{-1}\, S^2\, N(t) \\ &= R_0^{\frac{1}{k}}\, D(S^2 g, Sc \cdot SN(t))\, S(SN(t)), \end{aligned} \tag{3.13}$$

and the above stated parameter symmetry follows immediately. $\qquad\square$

An obvious consequence of this proposition is that bifurcation surfaces in (R_0, g, c)-parameter space occur k-fold, according to cyclic (g, c)-symmetry for fixed R_0.

4. The steady state

$L(I)$ is a Leslie matrix with eigenvectors

$$v_l = \begin{pmatrix} 1 \\ e^{-g_0 I}\,\lambda_l^{-1}\, R_0^{\frac{1}{k}} \\ e^{-(g_0+g_1) I}\,\lambda_l^{-2}\, R_0^{\frac{2}{k}} \\ \vdots \\ e^{-(g_0+\dots+g_{k-2}) I}\,\lambda_l^{-k+1}\, R_0^{\frac{k-1}{k}} \end{pmatrix} \tag{4.1}$$

and eigenvalues λ_l satisfying the characteristic equation

$$\lambda^k = R_0\, e^{-I}. \tag{4.2}$$

So a steady state requires, first of all,

$$\bar{I} = \ln R_0,\tag{4.3}$$

and then the steady state $\bar{N}$ is a multiple of the eigenvector

$$v_0 = \begin{pmatrix} 1 \\ e^{-g_0\,\bar{I}}\,R_0^{\frac{1}{k}} \\ e^{-(g_0+g_1)\,\bar{I}}\,R_0^{\frac{2}{k}} \\ \vdots \\ e^{-(g_0+\ldots+g_{k-2})\,\bar{I}}\,R_0^{\frac{k-1}{k}} \end{pmatrix}\tag{4.4}$$

such that $c \cdot \bar{N} = \bar{I}$. Hence the steady state is unique, lies in the interior of the positive cone, and is given explicitly by

$$\bar{N} = \frac{\ln R_0}{c \cdot v_0}\,v_0.\tag{4.5}$$

5. The linearized equation

To determine the linearized problem we put $N(t) = \bar{N} + x(t)$, which implies that $I(t) = \bar{I} + c \cdot x(t)$. We obtain the linearized equation

$$x(t+1) = L(\bar{I})x(t) + c \cdot x(t)\,R_0^{\frac{1}{k}}\,\mathrm{diag}\left(S(-g_0 e^{-g_0\bar{I}},\ldots,-g_{k-1}e^{-g_{k-1}\bar{I}})\right)S\,N.\tag{5.1}$$

To this equation we apply the transformation $x(t) = Ty(t)$ with

$$T = \mathrm{diag}\left(1, R_0^{\frac{1}{k}}\,e^{-g_0\,\bar{I}}, R_0^{\frac{2}{k}}\,e^{-(g_0+g_1)\,\bar{I}}\ldots, R_0^{\frac{k-1}{k}}\,e^{-(g_0+\ldots+g_{k-2})\,\bar{I}}\right).\tag{5.2}$$

This yields an equation for y where the shift mechanism is clearly visible:

$$y(t+1) = S\,y(t) + c \cdot Ty(t)\,w,\tag{5.3}$$

where

$$w = -\frac{\ln R_0}{c \cdot v_0}\begin{pmatrix} g_{k-1} \\ g_0 \\ \vdots \\ g_{k-2} \end{pmatrix} = -\frac{\ln R_0}{c \cdot v_0}\,Sg.\tag{5.4}$$

The *eigenvalue problem* is given by

$$S\,y + c \cdot T y\,w = \mu\,y.\tag{5.5}$$

Proposition 5.1. *The eigenvalues μ are the roots of the characteristic equation*

$$0 = \mu^k - 1 - \sum_{l=1}^{k}\mu^{k-l}\,c \cdot TS^{l-1}w.\tag{5.6}$$

Proof. We start from the eigenvalue equation

$$S\,y + c\cdot T y\, w = \mu\, y.$$

Applying S k-times and substituting $\mu\, y - c\cdot T y\, w$ for $\mu^j\, S y$ we obtain

$$S^k\, y = \mu^k y - \sum_{l=0}^{k-1} \mu^{k-l}\, c\cdot T y\, S^{l-1}\, w.$$

$S^k = \mathrm{Id}$, so taking the inner product with Tc we find

$$c\cdot T y = \mu^k\, c\cdot T y - \sum_{l=0}^{k-1} \mu^{k-l}\, c\cdot T y\, c\cdot T S^{l-1}\, w,$$

which we rewrite in the form

$$c\cdot T y \left(1 - \mu^k + \sum_{l=0}^{k-1} \mu^{k-l}\, c\cdot T S^{l-1}\, w\right) = 0.$$

If for none of the eigenvectors y, $c\cdot T y$ vanishes, then the result follows, because in that case the second factor needs to vanish, which is precisely the content of the proposition. To finish the proof we show that if $c\cdot T y$ vanishes for an eigenvector y, then the corresponding eigenvalue is a root of the equation (5.6) as well. Observe that $c\cdot T y = 0$ implies that y is an eigenvector of S (recall (5.5)). So the eigenvalue is simple and given by $\mu = \lambda_l = e^{2\pi i \frac{l}{k}}$, and the eigenvector y equals ξ_l where

$$\xi_l = \begin{pmatrix} 1 \\ e^{-2\pi i \frac{l}{k}} \\ \vdots \\ e^{-2\pi i \frac{l(k-1)}{k}} \end{pmatrix}. \tag{5.7}$$

So $c\cdot T y = 0$ can be rewritten as

$$c_0 + c_1 R_0^{\frac{1}{k}} e^{-g_0 \bar I} e^{-2\pi i \frac{l}{k}} + \ldots c_{k-1} R_0^{\frac{k-1}{k}} e^{-(g_0 + \ldots g_{k-2})\bar I} e^{-2\pi i \frac{l}{k}(k-1)} = 0. \tag{5.8}$$

We claim that consequently

$$\sum_{m=1}^{k} \lambda_l^{k-m}\, c\cdot T S^{m-1} w = 0. \tag{5.9}$$

To see this we fix j and compute the coefficient of g_j in the left-hand side of (5.9). Recall the form of w given in (5.4). When the summation variable in (5.9) equals m, then g_j is on the position $j + 1 + m \pmod{k}$ and is hence multiplied by

$$c_{j+m} R_0^{\frac{j+m}{k}} e^{-(g_0 + \ldots g_{j+m-1})\bar I} e^{-2\pi i \frac{l}{k} m}.$$

When we sum these terms over m they yield the left-hand side of (5.8) multiplied with $e^{2\pi i \frac{lj}{k}}$, which proves the claim and completes the proof. □

6. Equal sensitivity

In the special case of *equal sensitivity*, i.e.,

H1
$$g_0 = g_1 = \ldots = g_{k-1} = \frac{1}{k}, \tag{6.1}$$

equation (2.1) reduces to

$$N(t+1) = R_0^{\frac{1}{k}} e^{-\frac{c \cdot N(t)}{k}} S\, N(t). \tag{6.2}$$

In this case we have a *scalar nonlinearity*. In order to exploit this we shall represent the vectors $N(t)$ in terms of their *direction* $\sigma(t)$ (a unit vector, in the ℓ_1-sense, in the positive cone) and their *magnitude* $A(t)$. So define for positive k-vectors y the quantity

$$|y| := y_0 + \ldots + y_{k-1} \tag{6.3}$$

and put

$$N(t) = A(t)\,\sigma(t) \tag{6.4}$$

with the requirement that

$$|\sigma(t)| = 1. \tag{6.5}$$

Then, upon substitution into (6.2), we obtain the phase-amplitude equations

$$\begin{cases} \sigma(t+1) = S\,\sigma(t) \\[2mm] A(t+1) = R_0^{\frac{1}{k}} e^{-\frac{A(t)}{k} c \cdot \sigma(t)} A(t). \end{cases} \tag{6.6}$$

It follows that

$$\sigma(t+k) = \sigma(t). \tag{6.7}$$

In other words, every straight half-line through the origin is after k steps mapped onto itself. That is, we can decompose the collection of straight half-lines in the positive cone into a collection of invariant k-tuples of such half-lines, and the members of any k-tuple are mapped cyclically into each other. The half line spanned by the vector

$$\bar{\sigma} = \begin{pmatrix} \frac{1}{k} \\ \vdots \\ \frac{1}{k} \end{pmatrix} \tag{6.8}$$

is invariant. On this invariant half line we have amplitude dynamics

$$A(t+1) = R_0^{\frac{1}{k}} e^{-\frac{1}{k^2} A(t)} A(t) \tag{6.9}$$

with nontrivial fixed point

$$\bar{A} = k \ln R_0, \tag{6.10}$$

corresponding to the fixed point (4.5) which, in this special case, takes the form

$$\bar{N} = \bar{A}\,\bar{\sigma} = \begin{pmatrix} \ln R_0 \\ \vdots \\ \ln R_0 \end{pmatrix}. \tag{6.11}$$

At the "outer extreme" of the positive cone we have the k-tuple of positive half-axes (i.e., for some j we have $\sigma_l = 0$ for all $l \neq j$ and $\sigma_j = 1$) which are mapped cyclically into each other for all choices of g, not just in the present case of a uniform vector g.

We now prove the existence of an invariant manifold that contains the fixed point. To do so we introduce the vector $\mathbf{I}(t) \in \mathbb{R}^k$, whose first component is the scalar $I(t)$, see (2.4), while the other components are its iterates:

$$\mathbf{I}(t) = \begin{pmatrix} I_0(t) \\ I_1(t) \\ \vdots \\ I_{k-1}(t) \end{pmatrix} = \begin{pmatrix} c \cdot N(t) \\ c \cdot N(t+1) \\ \vdots \\ c \cdot N(t+k-1) \end{pmatrix}. \tag{6.12}$$

A condition is required for the transformation to be nonsingular. We omit the proof, which is straightforward.

Proposition 6.1. *The transformation from N-space to $\mathbf{I}$-space is nonsingular provided that*

H2
$$\begin{vmatrix} c_0 & c_1 & \cdots & c_{k-1} \\ c_1 & c_2 & \cdots & c_0 \\ \vdots & \vdots & & \vdots \\ c_{k-1} & c_0 & \cdots & c_{k-2} \end{vmatrix} \neq 0. \tag{6.13}$$

Remark 6.2. *This matrix is called a circulant. Its determinant is a homogeneous polynomial of degree k in the variables $c_0, \ldots, c_{k-1}$. There are no "simple" formulas* [Dav79].

Proposition 6.3. *Let H1-2 hold. The manifold*
$$\mathcal{M} = \{ \mathbf{I} \mid I_0 + \ldots + I_{k-1} = k \ln R_0 \}$$
is invariant under the mapping (2.1), and the restriction of the mapping (2.1) to $\mathcal{M}$ is, in $\mathbf{I}$-coordinates, represented by S^{-1}.

Proof. By definition we have $I_j(t+1) = I_{j+1}(t)$, for $j = 0 \ldots k-2$. We will show that the equality $I_{k-1}(t+1) = I_0(t)$ holds on $\mathcal{M}$, which proves the result.

$$I_0(t+1) = c \cdot N(t+1) = R_0^{\frac{1}{k}} e^{-\frac{I_0(t)}{k}} c \cdot S N(t). \tag{6.14}$$

Repeating this, i.e., increasing t in unit steps or, equivalently, increasing the lower index, we end up with

$$\begin{aligned} I_{k-1}(t+1) &= (R_0^{\frac{1}{k}})^k \, e^{-\frac{I_{k-1}(t) + \ldots + I_0(t)}{k}} \, c \cdot S^k N(t) \\ &= R_0 e^{-\frac{I_{k-1}(t) + \ldots + I_0(t)}{k}} \, c \cdot N(t) \\ &= I_0(t). \end{aligned} \tag{6.15}$$

$\square$

So we have found an invariant manifold, and on this manifold all points have period k. A point on $\mathcal{M}$ has I-coordinates $(I_0, \ldots, I_{k-1})$ and the mapping acts on this point as the shift S. In the next proposition we compute the multipliers of the k-th iterate of the full map, not just the restriction to $\mathcal{M}$, at this point.

Proposition 6.4. *Let H1-2 hold. The k-th iterate of the mapping (2.1) fixes every point of $\mathcal{M}$. The corresponding multipliers are 1, with multiplicity $k-1$, and*

$$\mu = \prod_{i=0}^{k-1} \left(1 - \frac{I_i}{k}\right), \tag{6.16}$$

with multiplicity 1.

Proof. The mapping takes in I-space the form

$$\begin{cases} I_0(t+1) = I_1(t) \\ \quad \vdots \\ I_{k-2}(t+1) = I_{k-1}(t) \\ I_{k-1}(t+1) = R_0\, e^{-\frac{I_0(t)+\ldots+I_{k-1}(t)}{k}}\, I_0(t). \end{cases} \tag{6.17}$$

On $\mathcal{M}$, the derivative of the k-th iterate is the product of k matrices

$$M = A_{I_{k-1}} \cdot A_{I_{k-2}} \cdot \ldots \cdot A_{I_0}, \tag{6.18}$$

where

$$A_a = \begin{pmatrix} 0 & 1 & & \\ & & \ddots & \\ & & & 1 \\ 1 - \frac{a}{k} & -\frac{a}{k} & & -\frac{a}{k} \end{pmatrix}$$

Observe that A_a has ξ_l, (5.7), as eigenvector, (independent of a) with eigenvalue λ_l for $l = 1 \ldots k-1$. The determinant of A_a is $(-1)^k (1 - \frac{a}{k})$. As $\lambda_l^k = 1$, 1 is an eigenvalue of M with multiplicity $k-1$, and the remaining eigenvalue is the product given in (6.16). $\qquad \square$

With this result at hand we can formulate a condition that guarantees the normal hyperbolicity of $\mathcal{M}$. We give a crude result that is an immediate consequence of the proposition above.

Proposition 6.5. *Let H1-2 hold. $\mathcal{M}$ is normally hyperbolic if $1 < R_0 < e^2$.*

Proof. We compute the extrema of μ subject to the conditions that $\mathbf{I} \in \mathcal{M}$ and $I_i \geq 0$, $i = 0 \ldots k-1$ (we suppress the argument t as it is here not relevant). We define

$$\mathcal{F} = \prod_{j=0}^{k-1} \left(1 - \frac{I_j}{k}\right) - \lambda \left(\sum_{j=0}^{k-1} I_j - k \ln R_0\right)$$

and we look for critical points of $\mathcal{F}$. This yields the set of equations

$$
\begin{cases}
\displaystyle\prod_{j=0}^{k-1}\left(1-\frac{I_j}{k}\right) = -k\lambda\left(1-\frac{I_i}{k}\right), & i = 0\ldots k-1 \\[4mm]
\displaystyle\sum_{j=0}^{k-1} I_j = k\ln R_0.
\end{cases}
\tag{6.19}
$$

Suppose $\lambda \neq 0$. The first equation of (6.19) tells us that I_i is independent of i. Then the second equation implies that $I_i = \ln R_0$, which yields $\mu = (1 - \frac{\ln R_0}{k})^k$.

If $\lambda = 0$, then at least two of the I_i should be equal to k and, accordingly, $\mu = 0$. So far we have not taken care of the constraint $I_i \geq 0$. We do that now by paying special attention to the situation where $k - l$ of the I_i are equal to 0, while the other ones are strictly positive. This amounts to redefining

$$
\mathcal{F} = \prod_{j=0}^{l-1}\left(1-\frac{I_j}{k}\right) - \lambda\left(\sum_{j=0}^{l-1} I_j - k\ln R_0\right)
$$

and repeating the analysis. We find $I_i = \frac{k}{l}\ln R_0$ and $\mu = (1 - \frac{\ln R_0}{l})^l$ which is inside the unit circle for $1 < R_0 < e^{2l}$. So the most severe constraint arises for $l = 1$. $\qquad\qquad\qquad\qquad\qquad\qquad\qquad\qquad\qquad\qquad\qquad\qquad\qquad\qquad\square$

References

[Beh00] H. Behncke. Periodical cicadas. *Journal of Mathematical Biology*, 40: 413–431, 2000.

[Bul77] M.G. Bulmer. Periodical insects. *The American Naturalist*, 111: 1099–1117, 1977.

[Dav79] P.J. Davis. *Circulant Matrices*. Pure and Applied Mathematics. John Wiley & Sons, 1979.

[DDvG] N.V. Davydova, O. Diekmann, and S.A. van Gils. Year class coexistence or competitive exclusion for strict biennials? Submitted.

Odo Diekmann   Stephan van Gils
Vakgroep Wiskunde   Faculty of Applied Mathematics
Utrecht University   University of Twente
Post-box 80010   Post-box 217
3508 TA Utrecht   7500 AE Enschede
The Netherlands   The Netherlands
O.Diekmann@math.uu.nl   s.a.vangils@math.utwente.nl

Trends in Mathematics:
Bifurcations, Symmetry and Patterns, 151–155
© 2003 Birkhäuser Verlag Basel/Switzerland

The Accumulation of Boundary Doubling for Modified Tent Maps

Paul Glendinning

Abstract. We describe the transition to chaos via boundary doubling for a particularly simple class of map. In this two parameter family of maps the accumulation of boundary doubling occurs on a curve in parameter space. We characterize this curve and use relate the form of this curve to a novel set of difference equations with proportional delay.

1. Introduction

The transition to chaos in the sense of positive topological entropy is one of the fundamental problems of applied dynamical systems. In [5] (see also [7]) we showed that if $f_\mu : I \to \mathbf{R}$ is a family of unimodal, or one hump, maps of the interval such that the image of the critical point of the map lies outside the interval for all relevant values of the parameter μ, then the transition to chaos can be via an infinite sequence of boundary bifurcations. At the boundary bifurcation, the boundary of the interval is periodic, creating periodic orbit of period 2^n. If these bifurcations occur at parameter values μ_n, then under some fairly weak assumptions about the maps being considered

$$|\mu_{n+1} - \mu_n| \to C|\mu_n - \mu_{n-1}|^2 \qquad (1)$$

as $n \to \infty$ where the constant C depends on the family of maps being considered. It was also pointed out that such maps might find application in models with some catastrophic breakdown threshold above which the model ceases to be a good description of the phenomenon. The simplest examples of these bifurcations arises in modified tent maps, and it is these which we consider in more detail below.

A tent map with threshold is a standard tent map [3, 8] with the added restriction that the map is undefined if $x > \mu$ for some $\mu \in (0,1)$, i.e.,

$$T_{\mu,s}(x) = \begin{cases} sx & \text{if } 0 \le x \le s^{-1}\mu \\ \text{undefined} & \text{if } s^{-1}\mu < x < s^{-1}(s-\mu) \\ s(1-x) & \text{if } s^{-1}(s-\mu) \le x \le \mu \end{cases} \qquad (2)$$

with $s > 1$ and $0 < \mu \le 1$. It is these maps which will be the focus of our attention in the remainder of this paper. Note that the boundary bifurcations will occur through the right-hand end point of the interval $[0, \mu]$, i.e., the point $x = \mu$ will become periodic.

The maps $T_{\mu,s}$ with the slight modification that $T_{\mu,s}(x) = \mu$ on the central interval, which makes the map a continuous map of the interval into itself, were considered briefly by Derrida, Gervois and Pomeau [4] in the context of period-doubling, and the renormalization argument of the next section can be found there. I suspect that little of what is contained here, except perhaps the last section, would be new to them. The special case of $s = 2$ is considered in [9] and related results can be found in [1].

2. Renormalization

If $\mu < s^{-1}(s - \mu)$ then the only recurrent dynamics in $[0, 1]$ is the fixed point at $x = 0$. As μ increases through $\mu_0 = s(s + 1)^{-1}$ a new fixed point, x_0 is created, and this fixed point has a unique preimage y_0 in $x < s^{-1}\mu$. An easy calculation yields

$$x_0 = s(s + 1)^{-1}, \qquad y_0 = (s + 1)^{-1} \tag{3}$$

If $\mu > \mu_0$ then we can consider the second iterate, T_1, of the map on the two intervals in $[y_0, x_0]$ on which it is well defined, and these branches of the second iterate map the intervals they are defined on into $[y_0, x_0]$ provided $\mu < \tilde{\mu}_1$ where

$$\tilde{\mu}_1 = \frac{s^2 + s - 1}{s(s + 1)}. \tag{4}$$

After an affine, orientation reversing change of variable so that the induced map is now defined on the interval $[0, 1]$ it is easy to show that if $\mu \in (\mu_0, \tilde{\mu}_1]$ then the induced map T_1 is again a map of the form (2), $T_{M,S}$, where

$$M = \frac{s - s(s + 1)(1 - \mu)}{s - 1} \quad \text{and} \quad S = s^2 \tag{5}$$

Note that if $\lim_{\mu \downarrow \mu_0}$ then $M = 0$ and if $\mu = \tilde{\mu}_1$ then $M = 1$ so the entire range of dynamics available to $T_{M,S}$ is realised by the induced map T_1. In particular, there is μ_1 at which an orbit of period one for the induced map (period two for the original map) is created by boundary bifurcation, and a new induced map T_2 (the second iterate of the second iterate of $T_{\mu,s}$) can be defined on $(\mu_1, \tilde{\mu}_2]$, where $\tilde{\mu}_2$ is the analogous parameter value to $\tilde{\mu}_1$.

We now use the standard bootstrap argument as is used in the case of period-doubling cascades. Set $\tilde{\mu}_0 = 1$. Then proceeding inductively we see that if $\mu \in [\mu_n, \tilde{\mu}_n]$ then the induced map T_n (made up of parts of the 2^nth iterate of T) is well defined on an interval which has a point x_{n-1} of period 2^{n-1} at one end point. If $\mu = \mu_n$, an orbit of period 2^n is created by boundary bifurcation, and if $\mu = \tilde{\mu}_n$ the two branches of the induced map stretch over the interval on which the map is defined and so, by standard arguments [8], the topological entropy of the original map is $2^{-n} \log 2$. The intervals $I_n = [\mu_n, \tilde{\mu}_n]$ form a nested sequence of closed intervals, and so

$$\mu_\infty = \bigcap_{n=0}^{\infty} I_n \tag{6}$$

is non-empty, and is the accumulation point of the two sequences, (μ_n) and $(\tilde{\mu}_n)$ (strictly speaking we have not shown that μ_∞ is a singleton). If $\mu = \mu_\infty$ then the renormalization process (the process of defining induced maps) can be repeated infinitely often. If $\mu < \mu_\infty$ then the only periodic orbits of $T_{\mu,s}$ have periods 2^n, $n = 0, 1, \ldots, N$, for some finite N and so the topological entropy of the map is zero, whilst if $\mu > \mu_\infty$, then $\mu > \tilde{\mu}_m$ for some m and so the entropy of the map is greater than or equal to $2^{-m} \log 2$.

3. Scaling

Equation (5) makes it relatively straightforward to verify the scaling, (1). We have already shown that $\mu_0 = s(s+1)^{-1}$. If T has parameter $\mu = \mu_n$, with μ_n defined in section 2, then the induced map T_n has parameters (M_n, S_n) with $M_n = S_n(S_n + 1)^{-1}$. Similarly, if T has $\mu = \mu_{n+1}$ then T_{n+1} has parameters (M_{n+1}, S_{n+1}) with $S_{n+1} = S_n^2$ by (5) and $M_{n+1} = S_{n+1}(S_{n+1} + 1)^{-1}$. Setting $\mu = M_n$, $s = S_n$ and $M = M_{n+1}$ in the first equation of (5) we find

$$M_n = \frac{S_n^2 + (S_n - 1)M_{n+1}}{S_n(S_n + 1)} \tag{7}$$

where we think of M_{n+1} as a function of S_n^2.

Now, at the parameter values μ_n, μ_{n+1} and μ_{n+2}, T_n is well defined and the corresponding parameters for this induced map are $M_{n,0}$, $M_{n,1}$ and $M_{n,2}$ where

$$M_{n,0} = \frac{S_n}{S_n + 1}, \quad M_{n,1} = \frac{S_n^2}{S_n^2 + 1}, \quad \text{and } M_{n,2} = \frac{S_n(S_n^3 - S_n + 1)}{S_n^4 + 1} \tag{8}$$

as is verified using the expression already given for μ_0 and two applications of (7). Now let

$$\Delta_{n,m} = M_{n,m+1} - M_{n,m} \tag{9}$$

and think of $\Delta_{n,m}$ as a function of $S_n = S_{n-1}^2$, cf. (11). Defining T_0 to be the original map (2) we see that equation (1) is equivalent to

$$\Delta_{0,n} \to C\Delta_{0,n-1}^2 \text{ as } n \to \infty \tag{10}$$

and it is this which we wish to demonstrate now.

If T_n has a boundary bifurcation which creates an orbit of period 2^m at $M_{n,m}$ then T_{n-1} has a boundary bifurcation creating an orbit of period 2^{m+1} at $M_{n-1,m+1}$ which can be obtained from $M_{n,m}$ using (7). Hence

$$\Delta_{n-1,m+1} = \frac{S_{n-1} - 1}{S_{n-1}(S_{n-1} + 1)} \Delta_{n,m}(S_{n-1}^2) \tag{11}$$

In other words, if n is large, so S_n is large, and $\Delta_{n,m} \sim S_n^{-r_{n,m}}$ then $\Delta_{n-1,m+1} \sim S_{n-1}^{-r_{n-1,m+1}}$ where $r_{n-1,m+1} = 2r_{n,m} + 1$, i.e.,

$$r_{n-k,n+k} = 2^k A - 1, \qquad \text{with} \ \ A = r_{n,m} + 1. \tag{12}$$

for $k = 0, \ldots, n$. A simple calculation using (8) shows that

$$\Delta_{n,1} = \frac{S_n(S_n - 1)}{(S_n^2 + 1)(S_n + 1)} \qquad \Delta_{n,2} = \frac{S_n(S_n - 1)^2}{(S_n^4 + 1)(S_n^2 + 1)} \tag{13}$$

so $r_{n,1} = 1$ and $r_{n,2} = 3$, and hence $r_{0,n+1} = 2.2^n - 1$ and $r_{0,n+2} = 4.2^n - 1$. Equation (10) now follows since

$$\Delta_{0,n+1}^2 \sim s^{-4.2^n + 1} \quad \text{and} \quad \Delta_{0,n+1}^2 \sim s^{-4.2^n + 2} = s.s^{-4.2^n + 1} \tag{14}$$

Note that a similar argument gives the same scaling result for the parameters $\tilde{\mu}_n$ and this, together with the remarks on entropy at the end of section two show that the entropy increases like the reciprocal of the logarithm of $|\tilde{\mu}_n - \mu_\infty|$, a result known to Derrida et al [4], see also [5, 9].

4. The accumulation curve

The accumulation of boundary doublings occurs at a parameter μ_∞ which is a function of s. The locus of this accumulation in the full two parameter space (μ, s) will be discussed here. On this locus, the map T_0 can be renormalized infinitely many times, and hence so can the induced map T_1. Hence, if the locus takes the form $\mu = F(s)$ then from (5) we find that

$$(s - 1)F(s^2) = s - s(s + 1)(1 - F(s)) \tag{15}$$

with $s > 0$. Note that $\lim_{s \to 1} F(s) = \frac{1}{2}$ and $\lim_{s \to \infty} F(s) = 1$ as we would expect. It is easier to approach (15) using the inverse of s, so if $G(s^{-1}) = F(s)$ and $t = s^{-1}$ then (15) may be rewritten as

$$G(t) = \frac{1}{1+t} + \frac{t(1-t)}{1+t}G(t^2) \tag{16}$$

Replacing t by t^2 throught to obtain an expression for $G(t^2)$ in terms of $G(t^4)$ and so

$$G(t) = \frac{1}{1+t} + \frac{t(1-t)}{(1+t)(1+t^2)} + \frac{t^3(1-t)(1-t^2)}{(1+t)(1+t^2)}G(t^4). \tag{17}$$

It is now relatively straightforward to repeat this process, giving

$$G(t) = \sum_{n=0}^{\infty} a_n(t)t^{2^n - 1} \tag{18}$$

where

$$a_0 = \frac{1}{1+t}, \quad a_1 = \frac{1-t}{(1+t)(1+t^2)}, \quad a_n = \frac{(1-t)\prod_{r=0}^{n-2}(1-t^{2^r})}{(1+t^{2^{n-1}})(1+t^{2^n})}, \quad n \geq 2. \tag{19}$$

An alternative approach to the solution of (16) is to pose a formal power series solution $G(t) = \sum_0^\infty y_n t^n$. After rewriting (16) as $(1+t)G(t) = 1 + t(1-t)G(t^2)$, substitution of this formal power series solution gives a linear differnece equation with proportional delay for the coefficients:

$$y_{2n+1} = -y_{2n} + y_n, \qquad y_{2n+2} = -y_{2n+1} - y_n \qquad (20)$$

with the initial condition $y_0 = 1$. These equations differ to those studied by Buhmann and Iserles [2, 6] for a discretized model of the pantograph equation on the boundary of stability of the trivial solution only in the sign of y_n in the second equation. Equations (20) have a curious self-similarity property. If $z_k = y_{k+1} + y_k$ then $z_{2n+1} = -z_{2n}$, so we need only consider the odd or the even z_k. Consider the even case and let $z_{2k} = b_k$. Then, noting that $z_{2n+2} + z_{2n} = z_n$ we find that the difference equation for b_k is precisely (20) – even the initial condition is the same! I have been unable to find a simple interpretation of this result, but the coefficients of the solutions to the discretized pantograph equations also have a fractal structure [6].

References

[1] K.M. Brucks, M. Misiurewicz and C. Tresser, *Monotonicity properties of the family of trapezoidal maps*, Comm. Math. Phys. **137** (1991), 1–12.

[2] M.D. Buhmann and A. Iserles, *On the dynamics of a discretized neutral equation*, IMA J. Num. Anal., **12** (1992), 339–363.

[3] P. Collet and J.P. Eckmann, *Iterated maps on the interval as dynamical systems*, Birkhauser, Boston, 1980.

[4] B. Derrida, A. Gervois and Y. Pomeau, *Universal metric properties of bifurcations of endomorphisms*, J. Phys. A, **12** (1978), 269–296.

[5] P. Glendinning, *Transitions to chaos in unimodal maps of the interval to the real line*, UMIST Appl. Math. Rep. preprint 00/2 (2000).

[6] A. Iserles, *Exact and discretized stability of the pantograph equation*, Appl. Numer. Math., **24** (1997), 295–308.

[7] H. Kokubu, M. Komuro and H. Oka, *Multiple homoclinic bifurcations from orbit-flip. I. Successive homoclinic doublings*, Int. J. Bif. & Chaos, **6** (1996), 833–850.

[8] S. van Strien, *Smooth dynamics on the interval*, New Directions in Dynamical Systems, ed. T. Bedford and J. Swift, Cambridge University Press, Cambidge, 1988.

[9] K. Zyczkowski and E.M. Bollt, *On the entropy devil's staircase in a family of gap-tent maps*, Physica D, **132** (1999), 392–410.

Paul Glendinning
Department of Mathematics
UMIST
P.O. Box 88
Manchester M60 1QD, U.K.
e-mail: p.a.glendinning@umist.ac.uk

Trends in Mathematics:
Bifurcations, Symmetry and Patterns, 157–165
© 2003 Birkhäuser Verlag Basel/Switzerland

Piecewise Rotations:
Bifurcations, Attractors and Symmetries

Arek Goetz and Miguel Mendes

Abstract. In this paper we investigate the most basic two-dimensional generalizations of interval exchange maps. The system studied is obtained by composing two rotations. We illustrate a new example of an attractor. The structure of this attractor appears to be present in the invertible piecewise rotation systems with two atoms. In the non-invertible case, we also illustrate a bifurcation mechanism leading to births of satellite systems.

1. Introduction

In this paper, we study planar systems of partially defined Euclidean rotations which are non-invertible. After defining our systems (section 2), we illustrate a theoretical model for perturbations of piecewise rotations which are called *8-attractors* leading to examples whose attractors contain an infinite number of discs (section 4 and 5). In section 6, we construct a new example of an attractor with an induced almost invertible dynamics. Finally, in section 7, we generalize basic concepts of symmetries and reversing-symmetries to the dynamics of piecewise rotations.

Our model involves the most basic examples of piecewise rotations with two domains separated by the discontinuity line. The rotation angles are fixed and perturbations are accomplished by varying one of the centers of rotations.

Piecewise rotations are examples of Euclidean piecewise isometries. These systems generalize well known and studied interval exchanges to a class of Euclidean two-dimensional piecewise isometries. Piecewise isometries appear in a variety of contexts and have been recently extensively studied as interval exchanges, interval translations, rectangular exchanges, polygonal and polyhedron exchanges and pseudogroup systems of rotations. Piecewise isometric maps appear naturally in billiards, dual billiards, theory of foliations and tilings. Piecewise rotations appear in the dynamics of so called digital filters. Detailed references can be found in Goetz [1998a].

2. Definitions

The phase space of piecewise rotations is the complex plane. We fix the point $S_0 \in \mathbb{R}^-$ and the angles of rotation $\alpha_0, \alpha_1 \in (0, 2\pi)$. Let T_0 be the rotation by α_0 about S_0 and let T_1 be the rotation by α_1 about some point S_1.

We define a family $\mathcal{T}$ of piecewise rotations $T : \mathbb{C} \to \mathbb{C}$ such that

$$Tx = \begin{cases} T_0 x \text{ if } x \in P_0 = \{x : Re\ x < 0\} \\ T_1 x \text{ if } x \in P_1 = \{x : Re\ x \geq 0\}. \end{cases} \tag{1}$$

The family $\mathcal{T}$ is parameterized by the position of the right center of rotation S_1 leaving the left center of rotation S_0 and the angles of rotations α_0 and α_1, fixed. The half planes P_0 and P_1 will be further called *atoms* of the map T.

3. 8-attractor system

The central theorem in Goetz [1998a] describes the dynamics of T when $S_1 \in \mathbb{R}^+$. In this case, there are two externally tangent discs: D_0 in the left atom (centered at S_0) and D_1 in the right atom (centered at S_1) which are fixed by the map (Fig. 1(left)). Discs are the trapping regions as once an orbit enters one of the discs, it never escapes from it. A surprising result is that all orbits are attracted to the fixed discs. However, there exist orbits that are never trapped in $D_0 \cup D_1$ – such orbits visit both atoms infinitely often and they accumulate on the union of two circles which are the boundaries of D_0 and D_1. The system also has strong attracting properties as entire neighborhoods of $D_0 \cup D_1$ are contracted to $D_0 \cup D_1$, that is $D_0 \cup D_1$ is a piecewise isometric attractor (Fig. 1(left)). The main result in Goetz [1998a] can be summarized as follows:

Theorem 3.1. *Suppose that $S_1 \in \mathbb{R}^+$. Then all orbits of T accumulate in the two maximal invariant discs $D_0 \cup D_1$ ($D_0 \subset P_0$ and $D_1 \subset P_1$) fixed by T (Fig. 1(left)). There exist orbits visiting both atoms infinitely often. The accumulation set of these orbits is the boundary of $D_0 \cup D_1$. Moreover, for all bounded sets $Y \supset D_0 \cup D_1$, the sets $T^n Y$ decrease to $D_0 \cup D_1$ as $n \to \infty$.*

Piecewise rotations $T \in \mathcal{T}$ described in Theorem 3.1 (or their restrictions to an invariant neighborhood of the fixed discs $D_0 \cup D_1$) will be further called *8-attractor maps* since the orbits of points not trapped in the 8-attractor, accumulate on a figure eight (see Fig. 1(left) once again).

The fixed discs D_0 and D_1 are maximal domains whose iterates are contained in exactly one atom. In general, it is convenient to describe maximal domains which are not broken up in to smaller pieces under the iteration of the map via symbolic description. The *coding map* $\sigma_T : \mathbb{C} \to \{0,1\}^{\mathbb{N}}$ encodes the forward orbit of a point $x \in \mathbb{C}$ by recording the indices of atoms visited by the orbit, $\sigma_T(x) = \omega_0 \omega_1 \ldots$, where $T^k x \in P_{\omega_k}$ and $\omega_k \in \{0,1\}$. A *cell* is the set of all points with the same coding.

4. 8-attractor emerges as a local map in a small perturbation

If the point S_1 lies off the real axis, then Theorem 3.1 no longer holds (compare Fig. 1(left) and 1(right)). If the perturbation of S_1 is small (the line joining S_0 and S_1 remains almost perpendicular to the imaginary axis), then some orbits

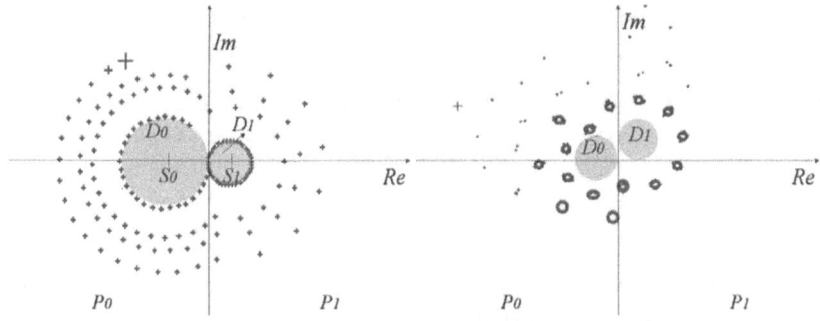

FIGURE 1. A single orbit (left figure) of an 8-attractor. All orbits are either trapped in one of the two shaded fixed discs, or they accumulate on the boundaries of the two fixed discs. If the right center of rotation is moved slightly off the real axis (right figure), for suitable choice of parameters, some orbits accumulate inside satellite periodic discs.

accumulate in a family of periodic discs and a new remarkable phenomenon of the birth of satellite systems is observed. The key starting idea in our model of this phenomenon is the next result (Theorem 4.1) which states that perturbations of S_1 can be always chosen so that a new periodic disc is externally tangent to D_0. Hence, a new 8-attractor system emerges locally in a small perturbation of an 8-attractor system. This gives rise to a new class of attractors, and also most importantly, the result can be recursively applied to the local map resulting in births of new satellite systems and systems with an infinite number of periodic discs.

In the following result, we perturb an 8-attractor map with centers of rotation $S_0 \in \mathbb{R}^-$ and $S_1 \in \mathbb{R}^+$ by varying S_1.

Theorem 4.1 (8-attractor emerges locally). *Every neighborhood of S_1 contains some $S_1' \in P_1$ such that:*

1. *The piecewise rotation $T \in \mathcal{T}$ with centers of rotation S_0 and S_1' has a periodic point $S_2 \in \mathbb{R}^+$ of some period l. The orbit of S_2 alternates between the atoms orbiting first around D_1' and then D_0. The cell containing S_2 is a periodic disc centered at S_2 and externally tangent to D_0.*

2. *A local map $T_\triangle$ defined as $T_\triangle x = Tx$ if $x \in P_0$, and $T_\triangle x = T^l x$ if $x \in P_1$ restricted to some neighborhood $M \supset D_0 \cup D_2$ is an 8-attractor map with centers of rotations S_0 and S_2.*

5. Recursive application of Theorem 4.1

Perturbations of 8-attractor maps frequently give rise to more than just one family of "satellite periodic discs" (see Fig. 6(b) in Goetz [1998b]). In this section we state that suitable, arbitrarily small perturbations of an 8-attractor map result in examples with infinitely many periodic cells which are discs.

Theorem 5.1 (Perturbations with infinitely many satellite discs). *The right center of every 8-attractor system can be perturbed in such a way that the new system has infinitely many periodic points. Every such periodic point p is contained in a cell which is a disc centered at p. Such a perturbation can be always chosen to be arbitrarily small.*

In order to illustrate the main idea of the proof of the above result note that Theorem 4.1 describes a perturbation G_1 of an 8-attractor map G_0 resulting in the birth of a small periodic disc D_2, orbiting around the fixed discs D_1' and then D_0. This theorem can be applied to a small perturbation of the local 8-attractor map $T_\triangle$. Since the position of the periodic point S_2 depends homeomorphically on the right center of rotation S_1, all small perturbations of $T_\triangle$ can be realized by suitable perturbations of S_1. In particular, there is a perturbation G_2 of G_1 with a small periodic disc D_3 and local map $T_{\triangle\triangle}$ restricted to a neighborhood of D_0 and D_3 is again an 8-attractor map. A new molecular attractor consisting of D_0 and the iterates of D_3 emerges. These perturbations can be repeated recursively. They converge (in the space of piecewise rotations) to a perturbation of an 8-attractor map with an infinite number of periodic discs.

6. Attractors

In the previous section we illustrated that for a suitable choice of irrational angles of rotations, the attracting set consists of an infinite number of periodic discs. Although in general the structure of an invariant set is unknown, in this section we illustrate that for a specific choice of rational angles of rotations, we can explicitly describe the attracting set and show that all orbits are trapped in the attractor. These new examples of attractors have complicated dynamics in their interior.

Details of the constructions and proofs can be found in Mendes [2001].

6.1. The notion of Attractor and Quasi-invariant sets

The next definition is a generalization of that presented in Goetz [1998a].

Definition 6.1. *A compact set A is called a* Piecewise Isometric Attractor *if there exists a neighbourhood $N \supset A$ such that $T(N) \subset N$ and*

$$A = \cap_{n \geq 0} T^n(N) \cup Z, \text{ for some set } Z \text{ such that } \mu(Z) = 0,$$

and also, such that there is no other compact set $A' \subset A$ satisfying the above equation for other zero measure set Z'.

Although these attractors may not be invariant sets, as all the existing examples prove, they might differ little from an invariant set as in the following sense,

Definition 6.2. *A positive measure set A is called* **quasi-invariant** *if $T(A) \subset A$ and $\mu(A \backslash T(A)) = 0$.*

For a reasonable large class of transformations we can derive a stronger property.

Proposition 6.3. *Let T be a mapping such that $\mu(T^{-1}(S)) = 0$ implies $\mu(S) = 0$ for all sets $S \subset Q$ where Q is a forward–invariant set. Then Q is quasi-invariant set if and only if there exists an invariant set $I \subset Q$ such that $\mu(Q \backslash I) = 0$.*

Moreover, we can show that the map T restricted to the attractor is *almost-invertible* in the sense that, $\mu(\{y \in T(A) : \#[T^{-1}(\{y\})] \neq 1\}) = 0$. More generally,

Proposition 6.4. *If T is a piecewise isometry and A is a quasi-invariant set then $T|_A$ is almost-invertible.*

This result generalizes the remark on Goetz [1998c] concerning a self-similar example which possesses rich dynamics for there exist infinitely many cells along with a fractal structure.

Furthermore, from above it follows that $T|_A$ is measure-preserving, and that every piecewise isometric attractor is a quasi-invariant set, since $\cap_{n \geq 0} T^n(N)$ is an invariant set.

The fact that some examples of attractors contain complex dynamical behaviour is closely related to the fact that they intersect the discontinuity line.

Finally, we pose some questions whose answer would give a deep insight to the classification of the possible attractors in piecewise rotations as well as which maps possess them.

Question: For which values of angles and centres of rotation do piecewise rotations generate attractors? For all those attractors containing fractal ω-limit sets, can we compute their Hausdorff dimension?

6.2. New examples

Our model is defined as follows: let $S_0 = (-1, -b)$ and $S_1 = (1, b)$, where b is some positive number; let $\alpha_0 = \alpha_1$ such that $T_0(\mathcal{D})$ and $T_1(\mathcal{D})$ are tangent to D_1 and D_0, respectively, where $\mathcal{D}$ is the discontinuity line. The quasi-invariant set is then contained in the area bounded by D_0, D_1 and the lines $T_0(\mathcal{D})$ and $T_1(\mathcal{D})$. In the following figure some phase portraits are depicted.

Theorem 6.5. *Take $\theta = \pi/n$, $n \geq 3$. Only for these choices of θ, the set described above satisfies the following:*

(a) $T(A) \subset A$ and $\mu(A \backslash T(A)) = 0$;
(b) *There exists an attracting neighbourhood N such that,*
   $A \supset \cap_{n=1}^{p_\theta} T^n(N)$ *for some $p_\theta \in \mathbb{N}$;*
(c) *Moreover, $A = \overline{\cap_{n=1}^{\infty} T^n(N)}$ and also $\mu(A \backslash \cap_{n=1}^{\infty} T^n(N)) = 0$.*

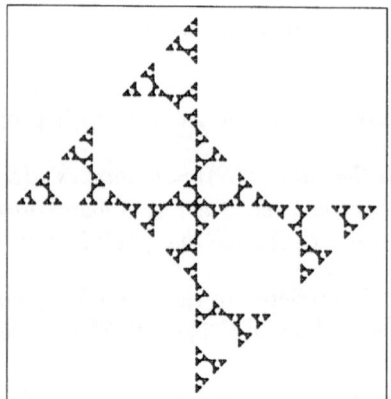 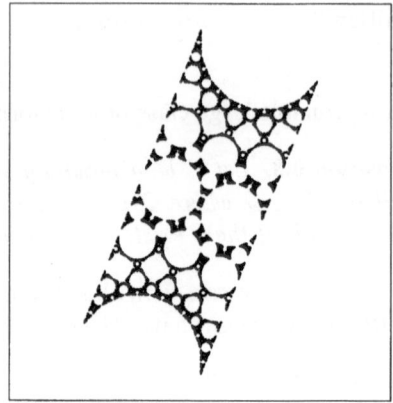

FIGURE 2. Fractals.

Finally, we state a result concerning a conjugacy between the maps described previously and models used in the theory of digital filters. These models are defined as follows,

$$x_{k+1} = \begin{bmatrix} 0 & 1 \\ -r^2 & 2r\cos\theta \end{bmatrix} x_k , \qquad x_k \in [-1,1[^2, \ \theta \in [0, 2\pi[$$

Proposition 6.6. *In the conditions of the previous theorem, the restriction of the map on the attractor is conjugated to the models presented above for $r = 1$.*

For the case when $\alpha = \pi/4$, a thorough study was carried out in Adler *et al.* [1999] and as well as in Ashwin *et al.* [1999], Guillaume [2000] and Kahng [2000]. Using self-similarities, the Hausdorff dimension of the fractal structure of the exceptional set (the black region) for $\alpha = \pi/4$ can be shown to be 1.246477....

7. Symmetries and Reversing–symmetries

In this section we introduce new concepts of symmetry that generalize the classic ones, as well as describe piecewise rotations that have a non-trivial group of symmetries for some subgroup of $\mathbb{D}_n < \mathbb{O}(n)$. Elements of $\mathbb{D}_n$ are assumed to act as reflections or rotations about the line of reflection or the centre of rotation. All proofs and more details of this section can be found in Mendes [2000].

The notions of *equivariance* and *reversibility* require that $f\sigma(x) = \sigma f(x)$ and $f\rho(x) = \rho f^{-1}(x)$, respectively, for all $x \in M$, where M is some manifold and $f : M \to M$ is a given mapping defined in M that generates a discrete dynamical system.

As far as piecewise rotations are concerned, we have found some classes that possess symmetries that resemble those from continuous maps. However, the former definitions would have neglected these new examples. Therefore, we will introduce weaker notions of equivariance and reversibility that suit discontinuous dynamical systems in general.

Definition 7.1. *Suppose σ is a linear isometry. Define $\Omega_\sigma = \{x \in M : f\sigma(x) = \sigma f(x)\}$. Then, σ is called an* **a.e.-symmetry** *if $\mu(M \backslash \Omega_\sigma) = 0$, where a.e. stands for almost everywhere.*

As usual, let $\mathcal{O}^+(x) = \{f^n(x) : n \in \mathbb{N}\}$ and define $\Omega_\sigma^n = \{x \in M : f^n\sigma(x) = \sigma f^n(x)\}$ and $\tilde{\Omega}_\sigma = \cap_{n \in \mathbb{N}} \Omega_\sigma^n$. We are interested on the effects that a.e.-symmetries might have on orbits. As stated below, a.e.-symmetries preserve a basic property in the classical setting.

Proposition 7.2. *Assume that f preserves sets of zero measure. Then $\mu(M \backslash \tilde{\Omega}_\sigma) = 0$ and in particular $\sigma(\mathcal{O}^+(x)) = \mathcal{O}^+(\sigma(x))$ for almost every x.*

In an analogous way, one defines *a.e.-reversing-symmetries*. From these definitions it is natural to say that f is *essentially-equivariant* if it possesses an a.e.-symmetry and *essentially-reversible* if it possesses an a.e.-reversing-symmetry as well as whether or not one can build a group of a.e.-(reversing)-symmetries, or shortly, group of symmetries.

Proposition 7.3. *The set of a.e.-(reversing)-symmetries form a group.*

In the next two theorems we describe all possible a.e-symmetries and a.e.-reversing-symmetries for piecewise rotations and, as a corollary, we present the resulting admissible symmetry groups.

Theorem 7.4. *If T is a piecewise rotation such that $\alpha_0 = -\alpha_1$ and $S_0 = -S_1 \in \mathbb{R}$ then T is essentially-equivariant for the a.e.-symmetry $\sigma.z = -\bar{z}$.*

If T is a piecewise rotation such that $\alpha_0 = \alpha_1$ and $S_0 = -S_1$ then T is essentially-equivariant for the a.e.-symmetry $\sigma.z = -z$.

Furthermore, there are no other cases of essential-equivariance with respect to $\mathbb{D}_n$, for piecewise rotations.

Let ρ_a be the reflection on the line passing through both centres of rotation and ρ_b the reflection on the line passing through the origin and perpendicular to the previous one.

Theorem 7.5. *Every invertible piecewise rotation is essentially-reversible for ρ_a. Furthermore, ρ_a and ρ_b are the only admissible a.e.-reversing-symmetries.*

Corollary 7.6. *The admissible groups of symmetry of piecewise rotations are:*

(a) $\mathbb{Z}_2$ *and,*
(b) $\mathbb{Z}_2 \times \mathbb{Z}_2$.

Case (a) includes maps with only one a.e.-reversing-symmetry and other maps where there is just one a.e.-symmetry. Whereas case (b), concerns those maps with a group of symmetry generated by an a.e.-symmetry and an a.e.-reversing-symmetry. Both of them generate normal subgroups isomorphic to $\mathbb{Z}_2$.

Question: Could this be generalized to piecewise rotations with more than two atoms?

In figure 3 we present some pattern-like phase portraits.

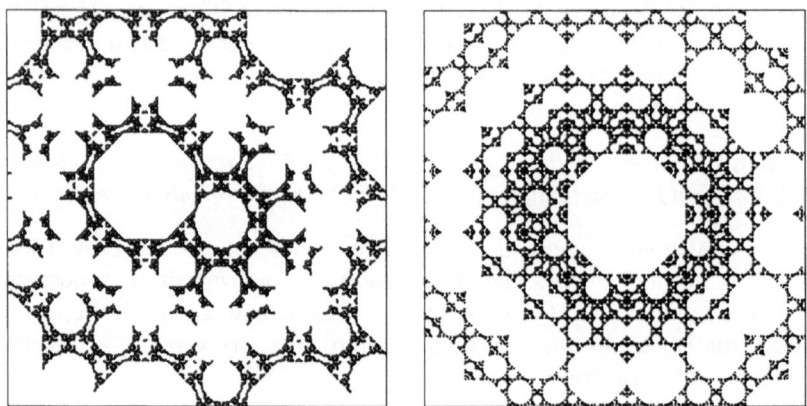

FIGURE 3. A purely essentially-reversible $\mathcal{PR}$ (left) and a $\mathbb{Z}_2 \times \mathbb{Z}_2$-essentially-reversible $\mathcal{PR}$ (right).

We shall now make two remarks concerning the examples of attractors in section 6.2. Firstly, those attractors have more symmetry, though we have not proved this, than the original piecewise rotations, a scenario which is not considered in the standard admissibility theory developed for homeomorphisms (see for instance, one of the introductory papers Ian Melbourne *et al* [1993]).

Secondly, the conjugacy with digital filters leads to the appearance of unusual symmetries as we shall now illustrate.

Assume that $f : A \to A$ and $g : B \to B$ are conjugated by $h : A \to B$, i.e., $hf = gh$. We define the push-forward of a (reversing)-symmetry σ of f as being the map $h^*(\sigma) = h\sigma h^{-1}$. It is easy to check that $h^*(\sigma)$ is a (reversing)-symmetry of g.

From the conjugacy mentioned previously in proposition 6.6, we can push-forward all existing (reversing)-symmetries from the original map that were proved in Adler *et al.* [1999]. Consequently, two new reversing-symmetries appear. However, they are piecewise isometric due to the fact that h is itself a piecewise isometry (h refers to the conjugacy established in proposition 6.6). We now give a more precise definition.

Definition 7.7. *Suppose σ is a piecewise linear isometry, i.e., there exist linear isometries σ_i and M_i for $i = 1, \ldots, p$ such that*

$$M_i \cap M_j \;=\; \emptyset, \; i \neq j \,, \; and \; M = \bigcup_{i=1,\ldots,p} M_i \,.$$

$$\sigma(x) \;=\; \sigma_i(x), \; if \; x \in M_i \,.$$

*Then, if $f\sigma(x) = \sigma f(x)$, for all $x \in M$, σ is called a **piecewise-symmetry**. If in addition, $\Omega_\sigma \neq \emptyset$ and $\mu(M\backslash\Omega_\sigma) = 0$, σ is called an **a.e.-piecewise-symmetry**. The same is defined for the reversible case.*

We hope that the symmetries that are present in our systems will help us prove the abundance of periodic points in general cases, which is a current focus of many researchers in the area.

References

[1] Adler, R., Kitchens, B. and Tresser, C. [1999] Dynamics of nonergodic piecewise affine maps of the torus, *preprint.*

[2] Ashwin, P., Chambers, W. and Petkov, G. [1997] Lossless digital filter overflow oscillations; approximation of invariant fractals. Inter. Journal of Bifur. and Chaos, Vol. **7** , No. 11 2603–2610.

[3] Goetz, A. [1998a] Dynamics of a piecewise rotation, Continuous and Discrete Dynamical Systems, 4 (4) 1998, p. 593–608. .

[4] Goetz, A. [1998b] Perturbations of 8-attractors and births of satellite systems, International Journal of Bifurcation and Chaos, Vol **8**, No. 10 1937–1956.

[5] Guillaume, P. [2000] Une isométrie par morceaux sur le tore, *Marseille University.*

[6] Kahng, B. [2000] Ph.D. Thesis, *University of Illinois at Urbana-Champain.*

[7] Melbourne, I., Dellnitz, M. and Golubitsky, M. [1993] The Structure of Symmetric Attractors, Arch. Rational Mech. Anal. **123** 75–98.

[8] Mendes, M. [2000] Piecewise Rotations and new concepts of Symmetry and Invariance, PhD Transfer Report, *University of Surrey.*

[9] Mendes, M. [2001] Quasi-Invariant Attractors of Piecewise Isometric Systems, *submitted for publication.*

Arek Goetz
Department of Mathematics
San Francisco State University
1600 Holloway Avenue
San Francisco, Ca 94080, U.S.A.
e-mail: goetz@math.sfsu.edu

Miguel Mendes
Department of Mathematics and Statistics
University of Surrey
GU2 7XH Guildford, U.K.
e-mail: m.mendes@surrey.ac.uk

Trends in Mathematics:
Bifurcations, Symmetry and Patterns, 167–174
© 2003 Birkhäuser Verlag Basel/Switzerland

Bound States of Asymmetric Hot Spots in Solid Flame Propagation

A. Bayliss, B.J. Matkowsky, A.P. Aldushin

1. Introduction

We consider modes of gasless solid fuel combustion, which are employed in the SHS (Self Propagating High Temperature Synthesis) process for the synthesis of advanced materials. In this process a finely ground powder mixture of desired reactants is ignited at one end. A high temperature thermal wave then propagates through the sample converting reactants to products. The SHS process was pioneered in the former Soviet Union and offers the promise of significant advantages in materials synthesis over conventional processes [9, 10].

Consider the burning of a cylindrical sample of solid fuel. The sample is ignited at one end and synthesis proceeds as a high temperature combustion wave propagates along the cylinder. In the simplest case the combustion wave has a uniformly propagating planar front separating the burned from the unburned sample, the temperature distribution along the front is uniform and the speed of the wave is constant. However, it is known both theoretically and experimentally that nonuniform modes of propagation are possible. Generally, these modes arise via bifurcations as parameters of the problem are varied. The study of different modes of propagation is significant since the nature of the combustion wave determines the conditions for synthesis, which affect the microstructure of the material produced. Indeed, some materials can only be synthesized in a nonuniform mode.

It has been observed that burning can occur throughout the sample, or can be confined to a narrow layer at the surface, though this is generally due to the effect of gas filtration. Nevertheless, we employ a gasless surface combustion model in which all radial dependence is neglected. The model also describes solid flame propagation in a thin cylindrical annulus between two concentric cylinders, as in the synthesis of hollow tubes. Thus, the independent variables are the axial coordinate $\tilde{z}$ and the cylindrical angle ψ. Planar modes are described by solutions which are independent of ψ.

A variety of nonuniformly propagating modes are known. These include planar pulsating combustion, sometimes referred to as autooscillatory combustion, in

Supported in part by NASA Grant NAG3-2209, NSF Grant DMS-0072491 and grant NWU 203 from the San Diego Supercomputer Center.

which planar fronts with a uniform temperature dependence on the front propagate with oscillatory velocities [11]. Such modes were described analytically by
bifurcation theory, as solutions of a mathematical model with $\delta-$function kinetics [8], and via numerical computations with Arrhenius kinetics, e.g., [3, 13]. The
transition from Arrhenius kinetics to δ-function kinetics occurs as the reaction
zone shrinks to a surface separating the burned and unburned regions. Nonplanar
modes of propagation typically occur when the radius of the cylinder is sufficiently
large. These modes are generally characterized by dynamics involving one or more
hot spots (localized high temperature maxima) along the combustion front. These
modes include (i) spin combustion, in which one or more hot spots move in a
helical fashion along the surface of the cylinder, all in the same direction, as the
combustion wave propagates, (ii) counterpropagating combustion in which spots
travel in opposite directions and undergo complex interactions when they collide,
e.g., apparent annihilation and creation, and (iii) multiple point combustion, in
which hot spot(s) repeatedly appear, disappear and reappear

Spinning waves were discovered experimentally in [11]. Such modes have also
been described analytically [7, 14] and numerically [1, 2, 5, 6]. Generally, information on the dynamics of the hot spots has been qualitative. Computations to
date have focused on spinning modes as symmetric traveling waves. Here, the term
symmetric means only that in the case of multiple hot spots, all spots are identical and are symmetrically placed, i.e., are equally spaced around the circle as
they propagate. The waves are not symmetric in shape, i.e., the spatial profiles of
the traveling pulses corresponding to the spots are asymmetric, as clearly seen in
Figure 4. We note that modulated single spot traveling waves have been observed
computationally, [1, 4].

In this paper we describe new types of spin combustion. Specifically, we
describe asymmetric traveling waves, in which two nonidentical spots, not equally
spaced around the circle, spin together as a traveling wave, or as a modulated
traveling wave. The leading spot has a higher temperature than the trailing spot,
thus generating more heat in the reaction to drive the wave. Nevertheless, the two
spots are bound together as they travel, due to heat transfer from the leading spot
to the trailing spot. Thus, the term bound states.

2. Mathematical model

The unknowns are the temperature T, and the mass fraction of a deficient component of the reactant $\tilde{Y}$. We consider a model which accounts for diffusion of
heat and one step, irreversible Arrhenius kinetics. Since the reactants are solid we
neglect diffusion of mass. Furthermore, we assume that burning is confined to the
surface of the cylindrical sample.

The nondimensional model in a coordinate system that moves with the front ϕ at velocity ϕ_t, is given by the system

$$\Theta_t = \phi_t \Theta_z + \Theta_{zz} + \frac{1}{R^2}\Theta_{\psi\psi} + ZY \exp\left(\frac{N(1-\sigma)(\Theta-1)}{\sigma+(1-\sigma)\Theta}\right),$$

$$Y_t = \phi_t Y_z - ZY \exp\left(\frac{N(1-\sigma)(\Theta-1)}{\sigma+(1-\sigma)\Theta}\right),$$

on the domain, $-\infty < z < \infty$, $0 \le \psi \le 2\pi$, $t > 0$, subject to the boundary conditions

$$\Theta(z,t) \to 0 \text{ as } z \to -\infty, \qquad \frac{\partial\Theta(z,t)}{\partial z} \to 0 \text{ as } z \to \infty,$$

$$Y(z,t) \to 1 \text{ as } z \to -\infty,$$

as well as periodicity in ψ and appropriate initial conditions. The boundary conditions in z are imposed at finite points as described in [1, 4]. Here, R is the radius of the cylinder (suitably nondimensionalized), N is nondimensionalized activation energy (typically large), $Z = N(1-\sigma)/2$ is the Zeldovich number ($Z \gg 1$), $\sigma = T_u/T_b$, where T_u and T_b are the temperature of the unburned mixture and adiabatic burning temperature respectively, and $Y = \tilde{Y}/\tilde{Y}_u$, $\Theta = (T-T_u)/(T_b-T_u)$, where $\tilde{Y}_u$ is the mass fraction of the deficient component far ahead of the combustion wave. The nondimensional independent variables are the time t, the cylindrical angle ψ, and the axial coordinate z in a reference frame moving with the front. Details on how the moving coordinate is obtained and further details on the model are given in [4], and details on the adaptive pseudo-spectral numerical method that we developed are given in [2]. The results presented here are obtained by varying N, keeping R and σ fixed. In all cases, we define the front location as the point at which the reaction rate is maximum.

3. Results

We first consider the evolution of asymmetric traveling waves, consisting of two nonidentical hot spots, not symmetrically located in angle, which spin together as a traveling wave. We call these modes   Bound States or Asymmetric Traveling Waves and refer to them as ATW2 solutions, indicating that there are two hot spots associated with these modes. These modes arise by bifurcation from two headed symmetric traveling waves (TW2), when the activation energy N increases above a critical value. The results indicate that they arise via a period doubling, symmetry breaking of the TW2 mode, as the symmetry of invariance with respect to rotation by angle π is broken. In Figures 1 and 2 we plot the temperature on the front for a fixed angular location as a function of t for a TW2 ($N = 22.75$) and for an ATW2 ($N = 23$) exhibiting the period doubling behavior as the ATW2 forms. In Figure 3 we exhibit a space-time plot of the temperature on the front as a function of ψ and t for an ATW2 solution. The straight ridges correspond to the hot spots, which spin at a uniform rate in a helical path around the cylinder. In Figure 4 we plot the

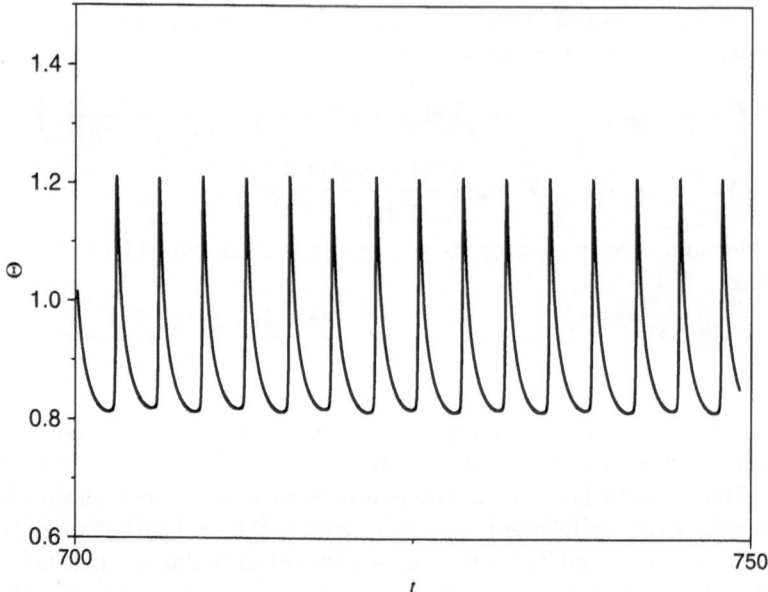

FIGURE 1. Θ at a fixed angular location on the front for a TW2 solution.

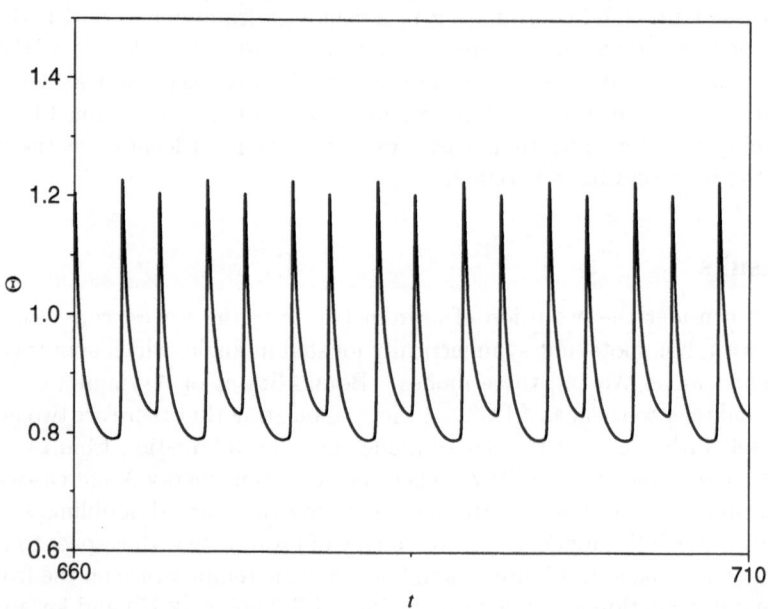

FIGURE 2. Θ at a fixed angular location on the front for an ATW2 solution.

FIGURE 3. Θ on the front as a function of ψ and t for an ATW2 solution.

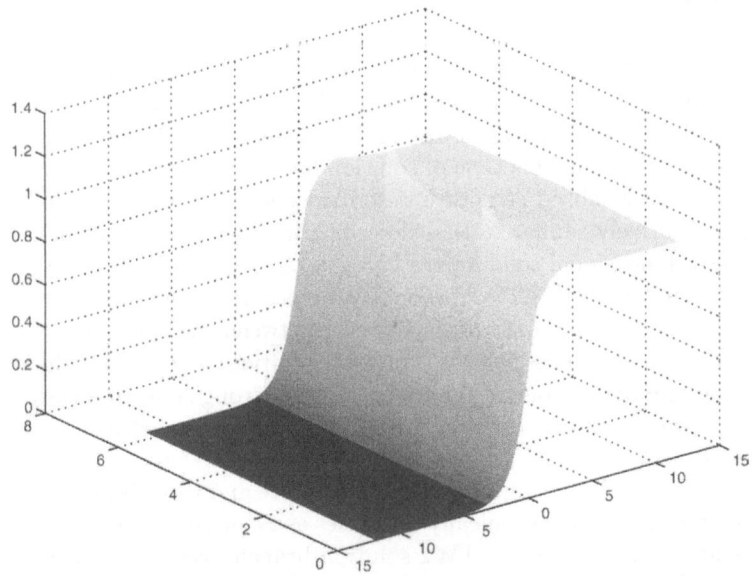

FIGURE 4. $\Theta(z, \psi)$ at a fixed instant of time for an ATW2 solution.

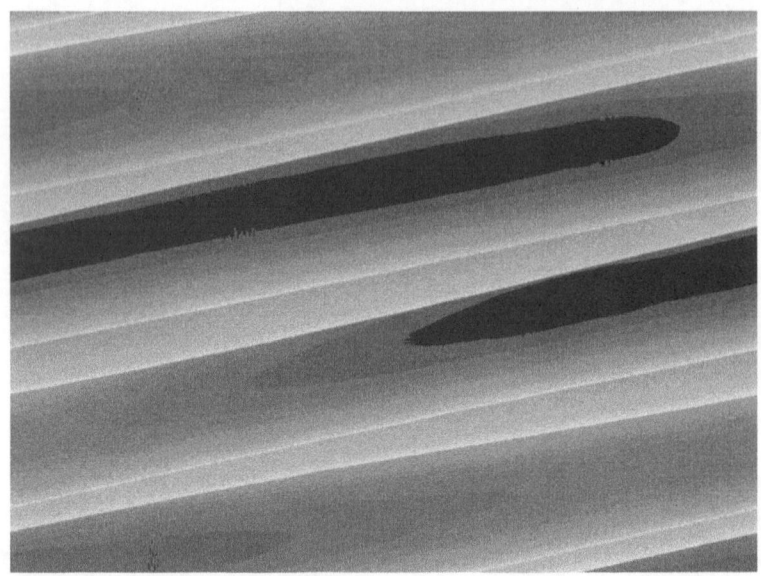

FIGURE 5. Θ on the front as a function of ψ and t for an AMTW2 solution.

temperature $\Theta(z, \psi)$ at a fixed instant of t, which shows a snapshot of the localized structure, which behaves as a kink in z and as a pulse in ψ. The characteristic feature of the ATW2 is that the spots are not identical and are not symmetrically spaced. A similar figure for the symmetric TW2 mode would show two identical spots separated by angle π. Thus, the ATW2 arises when the symmetry between the two spots is broken. This type of solution has been observed in one dimension [12], whereas the localized structure described here is two dimensional. A mode which is qualitatively similar to that described here has been described in a layer model of two dimensional solid flames [1].

We have followed the ATW2 branch by increasing N. In order to represent the branch we define a measure of the asymmetry between the two spots. Specifically, we define the quantity κ as follows. For each solution we compute the period T, i.e., the time required for one of the spots, say the strong spot, to return to a fixed angle. We also compute the smaller of the time differences between the arrival of two successive spots, δ. Then, for each solution $\kappa = \delta/T$. As $\kappa \to 0.5$, the solution approaches a TW2. As $\kappa \to 0$ the two spots coalesce and the solution approaches a TW1. We note that κ monotonically decreases as N increases, indicating that the two spots are coalescing. The ATW2 solution branch loses stability at a nonzero value of κ. When N increases beyond a critical value we find that the solution goes to a TW1, which is bistable with the ATW2 branch. Thus, we are unable to determine how the ATW2 branch emerges (perhaps as a branch of unstable solutions) from the TW1 branch. However, the ATW2 branch takes on some of the character of the TW1 branch as N increases, as illustrated by the monotonic decrease in κ. We note that for intermediate values of N we find tristability of the TW1, TW2 and ATW2 solutions. We have not determined whether the TW2 branch loses stability at the point of onset of the ATW2 branch.

Finally, we note that for larger values of R, the qualitative behavior between the TW2, ATW2 and TW1 branches is similar, except that upon losing stability we find a transition to asymmetric modulated traveling waves (AMTW2) in which the two asymmetric spots oscillate as they travel, first coming closer together and then spreading apart, as they propagate, with an accompanying oscillatory growth and shrinkage of the spots. In Figure 5 we illustrate such a solution in a space time plot over essentially one cycle of the apparently quasiperiodic motion.

References

[1] A.P. Aldushin, A. Bayliss, B.J. Matkowsky, *Dynamics of Layer Models of Solid Flame Propagation,* Physica D, to appear.

[2] A. Bayliss, R. Kuske and B. J. Matkowsky, *A two-dimensional adaptive pseudospectral method,* J. Comput. Phys. **91** (1990), 174–196.

[3] A. Bayliss and B. J. Matkowsky, *Two Routes to Chaos in Condensed Phase Combustion,* SIAM J. Appl. Math. **50** (1990) 437–459.

[4] A. Bayliss, B.J. Matkowsky, "Interaction of Counterpropagating Hot Spots in Solid Fuel Combustion", Physica D, **128** (1999), 18–40.

[5] T.P. Ivleva, A.G. Merzhanov, K.G. Shkadinsky, *Mathematical Model of Spin Combustion,* Soviet Physics Doklady, **23** (1978), 255–257.

[6] T.P. Ivleva, A.G. Merzhanov, K.G. Shkadinsky, *Principles of the Spin Mode of Combustion Front Propagation,* Combustion, Explosion and Shock Waves, **16** (1980), 133–139.

[7] S.B. Margolis, H.G. Kaper, G.K. Leaf, B.J. Matkowsky, *Bifurcation of Pulsating and Spinning Reaction Fronts in Condensed Two-Phase Combustion,* Combust. Sci. and Tech. **43** (1985), 127–165.

[8] B.J. Matkowsky, G.I. Sivashinsky, *Propagation of a Pulsating Reaction Front in Solid Fuel Combustion,* SIAM J. Appl. Math. **35** (1978), 465–478.

[9] A.G. Merzhanov, *SHS Processes: Combustion Theory and Practice,* Arch. Combustionis **1** (1981), 23–48.

[10] A.G. Merzhanov, *Self-Propagating High-Temperature Synthesis: Twenty Years of Search and Findings,* in: Combustion and Plasma Synthesis of High-Temperature Materials, Z.A. Munir, and J.B. Holt, Eds., VCH, (1990), 1–53.

[11] A.G. Merzhanov, A.K. Filonenko, I.P. Borovinskaya, *New Phenomena in Combustion of Condensed Waves,* Soviet Phys. Dokl. **208** (1973), 122–125.

[12] M. Or-Guil, I.G. Kevrekidis, M. Bär, "Stable Bound States of Pulses in an Excitable System", Physica D, **135** (1999), 157–174.

[13] K.G. Shkadinsky, B.I. Khaikin, A.G. Merzhanov, *Propagation of a Pulsating Exothermic Reaction Front in the Condensed Phase,* Combustion, Explosion and Shock Waves, **1** (1971), 15–22.

[14] G.I. Sivashinsky, *On Spinning Propagation of Combustion Waves,* SIAM J. Appl. Math., **40** (1981) 432–438.

A. Bayliss
Department of Engineering Sciences and Applied Mathematics
Northwestern University
Evanston, IL 60208

B.J. Matkowsky, A.P. Aldushin
Institute of Structural Macrokinetics and Materials Science
Russian Academy of Sciences
142432 Chernogolovka, Russia

Trends in Mathematics:
Bifurcations, Symmetry and Patterns, 175–180
© 2003 Birkhäuser Verlag Basel/Switzerland

Pattern Formation with Galilean Symmetry

P.C. Matthews and S.M. Cox

Abstract. The behaviour of pattern-forming systems in one dimension with Galilean symmetry in large domains is not described by the usual Ginzburg–Landau equation. This is because the Galilean symmetry leads to a large-scale neutral mode that interacts with the pattern. The resulting coupled amplitude equations, derived by considering the symmetry, show chaotic behaviour and exhibit a novel scaling in which the amplitude of the pattern is proportional to the 3/4 power of the bifurcation parameter.

1. Introduction

This paper is concerned with pattern formation in one dimension at a stationary bifurcation with non-zero wavenumber. Although this is the simplest type of pattern formation problem, some unanswered questions remain concerning the behaviour when the system is complicated by the presence of additional symmetries.

In finite domains, the spectrum of eigenvalues is discrete, and standard techniques of bifurcation theory with symmetry [5] can be used to show that the bifurcation is a pitchfork. In an infinite domain, the spectrum becomes continuous, so that a centre manifold reduction is not possible. In this case, the approach of applied mathematicians is to use a method of multiple scales, introducing a long lengthscale and a slow timescale based on the small parameter measuring the degree of supercriticality. This leads to the Ginzburg–Landau equation

$$A_T = A + A_{XX} - a|A|^2 A, \tag{1}$$

where a is a constant which can be scaled to ± 1. The crucial point about (1) is that it is generic for systems with Euclidean symmetry, provided that (i) the system has a uniform stationary state that becomes unstable at a stationary bifurcation at a finite wavenumber, and (ii) all other wavenumbers are damped and have growth rates bounded away from zero. Recently, this argument has been placed on a rigorous mathematical footing by Melbourne [9], who showed that a scalar equation is obtained, which when truncated at cubic order yields (1).

A key point in the derivation of (1) is that patterns with a wavenumber k generate modes with wavenumbers 0 and $2k$. These modes must be damped, and so slaved to the wavenumber-k modes, for (1) to be valid. However, there are a number of cases where the Ginzburg–Landau equation does not apply, due to the

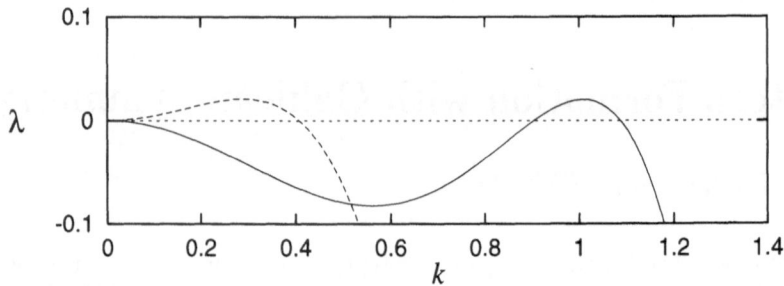

FIGURE 1. Spectrum of eigenvalues for systems with Galilean symmetry near the onset of pattern formation, for finite wavenumber (solid line) and small wavenumber (dashed line).

presence of some additional non-generic feature in the system. The wavenumber 0 is not damped (and therefore the result of [9] does not apply) if the system has a conservation law [8], or if there is Galilean symmetry. This paper will consider the latter case; a brief summary of some of the results has recently been published [7].

2. Galilean symmetry

Suppose that a PDE for $u(x,t)$ has Galilean and Euclidean symmetry,

$$x \rightarrow x + Vt + W, \qquad u \rightarrow u + V, \tag{2}$$

with the Euclidean reflection symmetry acting as -1, so that

$$x \rightarrow -x, \quad u \rightarrow -u. \tag{3}$$

These are very important symmetries in physics, possessed by many PDEs, for example the Navier–Stokes equations. Since the reflection symmetry acts as -1, u corresponds physically to a velocity.

For simplicity, we will suppose that the system is governed by a PDE that is first-order in time, so that

$$u_t = F(u, u_x, u_{xx}, \ldots). \tag{4}$$

Imposing that the equation is invariant under the symmetry (2) we find that $F_u + u_x = 0$ and hence

$$u_t = -uu_x + G(u_x, u_{xx}, \ldots). \tag{5}$$

Equivalently, one can calculate the invariants of (2), which are $u_t + uu_x$ and all x-derivatives of u, leading to (5). Thus the only linear terms permitted are terms involving derivatives of u.

The reflection symmetry (3) provides a further restriction on (5). In the linear terms, only even x-derivatives are permitted, so the eigenvalue λ of a mode proportional to $\exp(\lambda t + ikx)$ is of order k^2 as $k \rightarrow 0$.

There are now two possible scenarios for the onset of pattern formation. If the coefficient of the u_{xx} term is negative, then there is a long-wave instability and the small-amplitude behaviour is controlled by the Kuramoto–Sivashinsky equation, $u_t + uu_x = -u_{xx} - u_{xxxx}$. This paper is concerned with the alternative case, where the u_{xx} term is positive and the instability occurs at finite wavenumber. The spectrum of eigenvalues for these two cases is shown in Figure 1.

A simple model PDE satisfying the requirements of Galilean symmetry and bifurcation at a finite wavenumber is

$$\frac{\partial u}{\partial t} = -\frac{\partial^2}{\partial x^2}\left[ru - \left(1 + \frac{\partial^2}{\partial x^2}\right)^2 u\right] - u\frac{\partial u}{\partial x}, \tag{6}$$

which was proposed as a model for seismic waves [2]. The eigenvalues are given by

$$\lambda = k^2(r - (1 - k^2)^2), \tag{7}$$

so that instability occurs for $r > 0$, with critical wavenumber $k = 1$. Although (6) appears similar to the Swift–Hohenberg equation, its behaviour is quite different, and a number of studies of (6) have led to some controversy. It has been stated [10] that the amplitude of the solution is proportional to $r^{1/2}$ as $r \to 0$ (the usual scaling for a pitchfork bifurcation); but it has also been claimed [6] that the amplitude is proportional to r. In fact, as shown below, both of these scalings are incorrect.

3. Amplitude equations

We now derive appropriate amplitude equations for a PDE of the form (5), such as (6), with a stationary bifurcation at a non-zero wavenumber which we may take to be $k = 1$. The bifurcation is taken to occur when a control parameter r passes through zero, so that the growth rate λ is negative for $r < 0$ and positive for $r > 0$; r may be scaled so that $\lambda \sim r$ as $r \to 0$.

The Ginzburg–Landau equation is not valid, because the neutral large-scale mode corresponding to the limit $k \to 0$ must be taken into account as well as the mode with $k = 1$. The correct ansatz is

$$u = Re\left\{A(X,T)\exp(ix)\right\} + f(X,T), \tag{8}$$

where X and T are rescaled forms of x and t. The amplitude equations can be written down by requiring invariance under the symmetries of translation,

$$A \to A\exp(i\phi), \tag{9}$$

reflection,

$$X \to -X, \quad f \to -f, \quad A \to -A^*, \tag{10}$$

and Galilean symmetry

$$f \to f + V, \quad A \to A\exp(-iVT). \tag{11}$$

Applying these symmetries and scaling to remove constants leads to the equations

$$A_T = A + A_{XX} - a|A|^2 A - ifA, \tag{12}$$

$$f_T = \nu f_{XX} - \mu|A|^2{}_X, \tag{13}$$

where $a = \pm 1$, μ and ν are free parameters, and higher order terms, such as Af_X in (12), have been omitted. Alternatively, (12, 13) can be obtained by substituting (8) into (6), calculating the slaved $\exp(2ix)$ term and then equating the terms in $\exp(ix)$ and the terms independent of x.

Coullet and Fauve [3] were the first to write down equations similar to (12, 13); however they did not consider the scalings for A, f, X and T as $r \to 0$. If one adopts the standard scaling (used for the Swift–Hohenberg equation, or for nonlinear convection), $r = \epsilon^2$, $A = O(\epsilon)$, $X = \epsilon x$, $T = \epsilon^2 t$, then the amplitude equations are inconsistent, for any choice of the scaling for f. If, for example, $f = O(\epsilon^2)$ then (12) is consistent but ϵ^{-1} appears in the coupling term in (13), while if $f = O(\epsilon)$ then (13) is consistent but ϵ^{-1} multiplies the coupling term in (12). However, these inconsistent amplitude equations can be used to show that stationary patterns are always unstable, and that the growth rate of this instability is faster than the growth rate of the pattern.

The only way in which consistent amplitude equations can be obtained is by altering the scaling for A. The coupling terms balance if $r = \epsilon^2$, $A = O(\epsilon^{3/2})$, $f = O(\epsilon^2)$, $X = \epsilon x$, $T = \epsilon^2 t$, giving the asymptotically consistent equations

$$A_T = A + A_{XX} - ifA, \tag{14}$$

$$f_T = \nu f_{XX} - |A|^2{}_X, \tag{15}$$

where A has been rescaled so that $\mu = 1$. Note that the stabilising cubic term has been lost from (14). This allows exponentially growing solutions in which $A = A_0 \exp T$, but it can be shown that these are unstable, in the sense that spatially non-uniform perturbations to this state grow like $\exp 2T$ [7].

4. Numerical simulations

It is of interest to simulate numerically the model equation (6) in order to check the above novel scaling for the amplitude. This is a difficult computation, since to determine the correct scaling, r must be very small and the domain size L must be very large. Simulations were carried out using periodic boundary conditions with $L = 300$, using a spectral method with 512 grid points. Because of the six derivatives appearing, the system (6) is very stiff, so that special time-integration methods are required for efficient simulation. The exponential time differencing method [4] was used here; this allows much larger time steps than an explicit method but without the computational expense of an implicit method. The r.m.s. amplitude was averaged over a long time interval, for several values of r. Good agreement with the scaling law $u_{r.m.s.} \propto r^{3/4}$ was obtained over the range $0.001 < r < 0.1$, as shown in Figure 2.

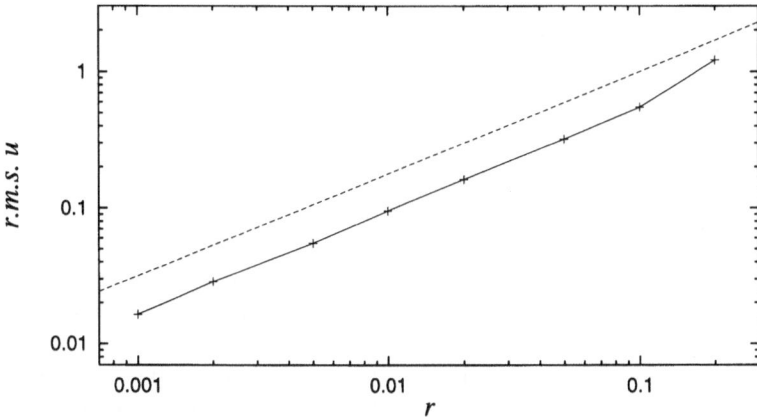

FIGURE 2. Log-log plot showing scaling of time-averaged r.m.s. amplitude with r for (6) with $L = 300$. Dashed line is of slope 3/4 for comparison.

A comparison between the behaviour of (6) and the amplitude equations (14, 15) with the appropriate parameters is shown in Figure 3. Note that the simulations of (6) show rapid destruction of the pattern and a subsequent evolution at lower amplitude, as suggested by the above scaling. The simulations of (14, 15) show a behaviour that is qualitatively very similar, and the amplitude remains bounded despite the lack of a stabilising cubic term in (14).

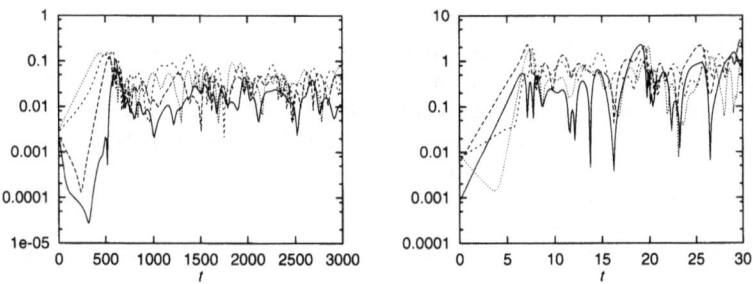

FIGURE 3. Numerical simulations of (6) (left, with $r = 0.01$, $L = 300$) and (14, 15) (right, with $\nu = 1/4$, $L = 15$), showing the evolution of four Fourier modes.

5. Discussion

In this paper we have derived the equations analogous to the Ginzburg–Landau equation for a stationary bifurcation at a finite wavenumber with Galilean symmetry. The Galilean symmetry leads to the existence of a neutral large-scale mode that is strongly coupled to the finite-wavenumber mode. Because of this interaction, there are no stable patterns, the behaviour is chaotic at onset and the amplitude scales as the 3/4 power of the bifurcation parameter.

There are a number of complications with regard to applying these results to experimental systems. Firstly, it is not possible to obtain true Galilean symmetry in an experiment. For example, although the Navier–Stokes equations have Galilean symmetry, the symmetry is weakly broken (by an amount proportional to the viscosity) by the presence of rigid boundaries. Secondly, in most applications [1, 2] there is no reflection symmetry, allowing additional terms in the amplitude equations. Further work is in progress on these problems.

References

[1] P. Barthelet and F. Charru (1998) Benjamin–Feir and Eckhaus instabilities with Galilean invariance: the case of interfacial waves in viscous shear flows. Eur. J. Mech. B **17**, 1–18.

[2] I.A. Beresnov and V.N. Nikolaevskiy (1993) A model for nonlinear seismic waves in a medium with instability. Physica D **66**, 1–6.

[3] P. Coullet and S. Fauve (1985) Propagative phase dynamics for systems with Galilean invariance. Phys. Rev. Lett. **55**, 2857–2859.

[4] S. M. Cox and P. C. Matthews (2001) Exponential time differencing for stiff systems. J. Comp. Physics, submitted.

[5] M. Golubitsky, I. Stewart and D.A. Schaeffer (1988) *Singularities and groups in bifurcation theory. Volume II.* Appl. Math. Sci. **69**, Springer-Verlag, New York.

[6] I.L. Kliakhandler and B.A. Malomed (1997) Short-wavelength instability in presence of a zero mode: anomalous growth law. Phys. Lett. A **231**, 191–194.

[7] P.C. Matthews and S.M. Cox (2000) One-dimensional pattern formation with Galilean invariance near a stationary bifurcation. Phys. Rev. E **62**, R1473-R1476.

[8] P.C. Matthews and S.M. Cox (2000) Pattern formation with a conservation law. Nonlinearity **13**, 1293–1320.

[9] I. Melbourne (1999) Steady state bifurcation with Euclidean symmetry. Trans. Amer. Math. Soc. **351**, 1575–1603.

[10] M.I. Tribelsky and K. Tsuboi (1996) New scenario for transition to turbulence? Phys. Rev. Lett. **76**, 1631–1634.

P.C. Matthews and S.M. Cox
School of Mathematical Sciences
University of Nottingham, Nottingham NG7 2RD, UK
e-mail: `paul.matthews@nottingham.ac.uk`
e-mail: `stephen.cox@nottingham.ac.uk`

Trends in Mathematics:
Bifurcations, Symmetry and Patterns, 181–187
© 2003 Birkhäuser Verlag Basel/Switzerland

Semigroups of Functions and the Structure of Stationary Measures in Systems which Contract-on-average

Matthew Nicol, Nikita Sidorov, and David Broomhead

1. Introduction

This note is a summary of a talk given at the Conference on Symmetry and Bifurcation in honor of Marty Golubitsky and Ian Stewart. The results stated in this note are found, together with proofs, in the paper [8].

Suppose $\{f_1, \ldots, f_m\}$ is a set of Lipschitz maps of $\mathbb{R}^d$. We form an iterated function system (IFS) by independently choosing the maps so that the map f_i is chosen with probability p_i ($\sum_{i=1}^m p_i = 1$). We assume that the IFS contracts-on-average (defined below). Such systems possess stationary invariant measures. We will describe recent work of Broomhead, Nicol and Sidorov [8] on the relation between the Hausdorff dimension of the measure, the entropy and the Lyapunov exponent of the system and the semigroup generated by the functions $\{f_1, \ldots, f_m\}$. We include several examples with a view to illustrate the role of the semigroup generated by the mappings.

To simplify the notation we will denote the probability vector by $\bar{p} := (p_1, \ldots, p_m)$, and the IFS itself by Φ. Another way of viewing the IFS is to let $\Omega = \prod_0^\infty \{1, \ldots, m\}$ and equip Ω with the product probability measure ν induced in the standard way on cylinder sets by the probability vector $\bar{p}$. Let $x_0 \in \mathbb{R}^d$. For any $\omega \in \Omega$ and any $n \in \mathbb{N}$ we may define the point

$$x_n(\omega) := f_{\omega_0} \ldots f_{\omega_{n-1}}(x_0).$$

If $\lim_{n \to \infty} x_n(\omega)$ exists, then we define

$$\phi(\omega) = \lim_{n \to \infty} x_n(\omega). \tag{1.1}$$

We will now describe conditions (the contraction-on-average condition) that ensures that $\phi(\omega)$ is defined μ a.e. independently of x_0.

Let

$$h(\bar{p}) := \sum_{i=1}^m p_i \log p_i.$$

where log is the natural logarithm. $-h(\bar{p})$ is the measure-theoretic entropy of the Bernoulli shift $\sigma : \Omega \to \Omega$ with the probabilities $(p_1, \ldots, p_m)$.

For any Lipschitz map g of $\mathbb{R}^d$ we let $\|g\|$ denote the Lipschitz constant of g. We assume a *contraction-on-average* (sometimes called *logarithmic average contractivity*) condition to hold: namely for ν-a.e. $\omega \in \Omega$,

$$\lim_{n \to \infty} \frac{1}{n} \log^+ \|f_{\omega_0} \cdots f_{\omega_{n-1}}\| = \chi(\Phi) < 0. \tag{1.2}$$

We call $\chi(\Phi)$ the *Lyapunov exponent* of the system.

Note that the condition (1.2) is implied by the easy to check condition

$$\sum_{i=1}^{m} p_i \log \|f_i\| < 0.$$

Under these assumptions it is well known (see P. Diaconis and D. Freedman [3]) that the IFS possesses a unique *stationary probability measure* μ on $\mathbb{R}^d$ independent of the choice of initial point. In fact the measurable mapping ϕ induces μ on Borel sets of $\mathbb{R}^d$ by $\mu(B) = \nu \circ \phi^{-1}(B)$. Sometimes we will also call μ the *invariant measure*.

By results of L. Dubins and D. Freedman [4] (see also M. Barnsley and J. Elton [2, Proposition 1]) on Markov operators, μ must be of *pure type*, i.e., either absolutely continuous or purely singular with respect to Lebesgue measure on $\mathbb{R}^d$. M. Barnsley and J. Elton [2, Theorem 3] show that if $d = 1$ and the maps $\{f_i\}$ are affine, i.e., $f_i(x) = \lambda_i x + \alpha_i$ and $|\lambda_i| \geq 1$ for at least one i then the support of μ is either $\mathbb{R}$ or of form $[a, \infty)$ or $(-\infty, b]$ for $a, b \in \mathbb{R}$. We exploit this observation in Section 3.

Example 1.1. An important classical example of an IFS which contracts strictly is the one-parameter family

$$f_0(x) = \lambda^{-1} x,$$
$$f_1(x) = \lambda^{-1} x + 1 - \lambda^{-1}$$

with $p_1 \in (0, 1)$ and $|\lambda| > 1$. It has been extensively studied since the 1930's. In recent work by B. Solomyak [11] it was shown that if $p_1 = p_2 = \frac{1}{2}$, then a.e. $\lambda \in (1, 2)$ induces an absolutely continuous measure μ on the interval $[0, 1]$. A similar result was later obtained by the same authors for $p_1 \in [1/3, 2/3]$ (see Section 3). The general problem of finding for which values λ and p this system is absolutely continuous or singular is very hard and only few concrete results are known (see [9] for a nice review and collection of references).

Recall that the *(upper) Hausdorff dimension* of a probability measure μ is the infimum of the Hausdorff dimensions of Borel sets B such that $\mu(B) = 1$. We denote the upper Hausdorff dimension of μ by $\dim_H(\mu)$.

In [8] the authors give an upper bound for the upper Hausdorff dimension of the invariant measure μ and describe sufficient conditions for μ to be singular in terms of $\chi(\Phi)$, $h(\bar{p})$ and the expansion rate of the semigroup generated by $\{f_i\}$. In Section 3 we present several examples showing the role of symmetry and how to apply the main theorem. In particular, for any $d > 2$ we give examples of singular measures whose support is the whole of $\mathbb{R}^d$ (these observations are additional to the cases $d = 1, 2$ of [8]). The key theme is that the problem of estimating $\dim_H(\mu)$ may be reduced to certain combinatorial and algebraic issues concerning the semigroup generated by the maps of the IFS.

2. Sufficient conditions for singularity of the invariant measure

Let G^+ denote the semigroup generated by the maps $\{f_1, \ldots, f_m\}$. Its elements are all compositions $f_{\omega_0} \circ \cdots \circ f_{\omega_{n-1}}$ for any $n \in \mathbb{N}$ and $\omega_k \in \{1, \ldots, m\}$. G^+ can be either the free semigroup $\mathcal{F}_m^+$ (if all such compositions are different) or a proper subsemigroup of $\mathcal{F}_m^+$.

Let D_n denote the set of all words of length n in G^+. In other words, D_n is the set of equivalence classes in $\prod_0^{n-1}\{1, \ldots, m\}$, namely:

$$(\omega_0^*, \ldots, \omega_{n-1}^*) \sim (\omega_0', \ldots, \omega_{n-1}') \text{ if } f_{\omega_0^*} \circ \cdots \circ f_{\omega_{n-1}^*} = f_{\omega_0'} \circ \cdots \circ f_{\omega_{n-1}'}.$$

By a standard argument there exists $\theta \in [1, m]$ such that

$$\theta = \lim_{n \to +\infty} \sqrt[n]{\#D_n}. \tag{2.3}$$

If G^+ is abelian, then $\theta = 1$ while if there are no relations between the mappings then $\theta = m$.

A main result of [8] is:

Theorem 2.1. *If*

$$\frac{h(\bar{p}) \log \theta}{\chi(\Phi)} < d \log m,$$

then μ is singular, and

$$\dim_H(\mu) \leq \frac{h(\bar{p}) \log \theta}{\chi(\Phi) \log m} < d.$$

If $p_1 = \cdots = p_m = \frac{1}{m}$, then

$$\dim_H(\mu) \leq \frac{\log \theta}{|\chi(\Phi)|}. \tag{2.4}$$

This has the following immediate corollaries:

Corollary 2.2. *If $\theta = 1$ (i.e., the semigroup G^+ grows subexponentially), then μ is singular and $\dim_H(\mu) = 0$.*

Corollary 2.3. *The measure μ is singular for any Φ such that*

$$d|\chi(\Phi)| > |h(\bar{p})|. \tag{2.5}$$

If (2.5) holds, then μ is singular and

$$\dim_H(\mu) \leq \frac{h(\bar{p})}{\chi(\Phi)} < d.$$

3. Examples

We are going to consider several examples to illustrate these results. All the examples we give will involve affine maps.

Example 3.1. We start with a simple example. Suppose $f_1(x) = 2x + 1$, $f_2(x) = \frac{1}{16}x + 1$ chosen with probabilities $p_1 = p_2 = \frac{1}{2}$. Then $\chi(\Phi) = -\frac{3}{2}\log 2 < h(\overline{p}) = -\log 2$ and hence by Corollary 2.3, the invariant measure μ is singular with respect to Lebesgue measure, and $\dim_H(\mu) \leq \frac{2}{3}$. However it is easy to show that the topological support of the invariant measure is the interval $[1, \infty)$. Note that a more detailed analysis shows that since $f_1 f_2 f_1^3 f_2 f_1 = f_2 f_1^5 f_2$, we have $\theta < 1.9836$, whence by the formula (2.4), $\dim_H(\mu) \leq \frac{2}{3}\log_2 \theta < 0.6588$.

Example 3.2. (Bernoulli convolutions). Put $\Omega := \prod_0^\infty \{0, 1\}$ and let $\lambda > 1$, $f_0(x) = \lambda^{-1}x$, $f_1(x) = \lambda^{-1}x + 1 - \lambda^{-1}$, $p_1 = p_2 = \frac{1}{2}$ (see Introduction). In this case $\chi(\Phi) = -\log \lambda$, and

$$f_{\omega_0} \circ \cdots \circ f_{\omega_{n-1}}(0) = (1 - \lambda^{-1}) \sum_{k=0}^{n-1} \omega_k \lambda^{-k},$$

thus,

$$\phi(\omega) = (1 - \lambda^{-1}) \sum_{k=0}^{\infty} \omega_k \lambda^{-k}.$$

The easiest subcase is $\lambda = \frac{1+\sqrt{5}}{2}$. It was studied in several papers (see references in [10]); in particular, for this λ we have $G^+ = \langle a, b \mid ab^2 = ba^2 \rangle$ and $\dim_H(\mu) = 0.995713\ldots$ (this numerical result is due to J. C. Alexander and D. Zagier [1]).

Example 3.3. Let $\lambda > 1$ and

$$f_1(x) = \lambda^{-1}x, \quad f_2(x) = x + 1 \quad \text{and} \quad p_1 = p_2 = \frac{1}{2}. \tag{3.6}$$

The support of $\mu = \mu(\lambda)$ is $[0, +\infty)$, and $\chi(\Phi) = -\frac{1}{2}\log \lambda$. Hence by Corollary 2.3, for $\lambda > 4$ the measure μ is singular, and $\dim_H(\mu) \leq \frac{2\log 2}{\log \lambda} < 1$.

We claim that for any transcendental λ the semigroup G^+ is free. An induction argument shows that

$$f_1^{n_1} f_2^{k_1} \cdots f_1^{n_s} f_2^{k_s}(x) = \lambda^{-\sum_1^s n_j} x + \sum_{j=1}^{s} k_j \lambda^{-\sum_{i=1}^{j} n_i},$$

so if λ is not algebraic,

$$f_1^{n_1} f_2^{k_1} \cdots f_1^{n_s} f_2^{k_s} = f_1^{n_1'} f_2^{k_1'} \cdots f_1^{n_s'} f_2^{k_s'}$$

implies $n_j \equiv n_j'$, $k_j \equiv k_j'$, $j = 1, \ldots, s$. Hence $\#D_n = 2^n$, $\theta = 2$ and $G^+ = \mathcal{F}_2^+$. Nevertheless there exist values of $\lambda \in (1, 4)$ for which μ is singular. For example, in [8, Example 3.2] it is shown that if $\lambda = \frac{(1+\sqrt{5})}{2}$ then μ is singular.

As far as we know, there are no general results on the structure of $\mathrm{supp}(\mu)$ in the case of higher dimensions. We generalise an example in [8] to present an example of a family of IFS for $d > 2$ such that $\mathrm{supp}(\mu) = \mathbb{R}^d$, whereas μ is singular.

Example 3.4. We now generalise an example in [8] from $d = 2$ to $d > 2$. In [8] a measure is constructed whose support is the whole of the plane, yet the measure is singular. This construction proceeds as follows: let $\lambda > 1$ and the one-parameter family of IFS Φ_λ be defined by

$$f_1(x) = A_\lambda^{-1}x,$$
$$f_2(x) = x + \bar{e},$$

where $A_\lambda = \lambda R_\alpha$, α/π is irrational, $\bar{e} \in S^1$ and

$$R_\alpha = \begin{pmatrix} \cos\alpha & -\sin\alpha \\ \sin\alpha & \cos\alpha \end{pmatrix}.$$

In [8] it is shown that:

Proposition 3.5. [8]

1. *For any $\lambda > 1$ the IFS Φ_λ contracts on average and the support of the invariant measure μ_λ is full, i.e.,*

$$supp(\mu_\lambda) = \mathbb{R}^2. \tag{3.7}$$

2. *For $\lambda > 2$ the measure μ_λ is singular, and*

$$\dim_H(\mu_\lambda) < \frac{2\log 2}{\log\lambda} < 2. \tag{3.8}$$

This construction has a natural generalization to $\mathbb{R}^d$. The main difference is that although $\mathbf{SO}(2)$ needs just one generator, if $d > 2$ then $\mathbf{SO}(d)$ requires two generators. In fact $\mathbf{SO}(d)$ has a right and left invariant Haar measure h and $h \times h$ a.e. pair of elements (γ_1, γ_2) generate $\mathbf{SO}(d)[7, 6]$. Choose two such generators (γ_1, γ_2) with matrix representations M_1, M_2. As above let $\lambda > 1$ and define the one-parameter family of IFS Φ_λ as follows:

$$f_1(x) = M_{1,\lambda}^{-1}x,$$
$$f_2(x) = M_{2,\lambda}^{-1}x,$$
$$f_3(x) = x + \bar{e},$$

where $M_{i,\lambda} = \lambda M_i, (i = 1, 2)$ and $\bar{e} \in \mathbb{R}^d$ is fixed and of unit length (ie $\|\bar{e}\| = 1$). We assume $p_1 = p_2 = \frac{1}{4}$ and $p_3 = \frac{1}{2}$. The analogous statement is:

Proposition 3.6. 1. *For any $\lambda > 1$ the IFS Φ_λ contracts on average and the support of the invariant measure μ_λ is full, i.e.,*

$$supp(\mu_\lambda) = \mathbb{R}^d. \tag{3.9}$$

2. *For $\lambda > 2^{\frac{3}{d}}$ the measure μ_λ is singular, and*

$$\dim_H(\mu_\lambda) < \frac{3\log 2}{\log \lambda} < d. \tag{3.10}$$

Proof. Corollary 2.3 implies that $\lambda > 2^{\frac{3}{d}}$ ensures the singularity of μ_λ together with (3.10).

We now prove 3.9. Assume $\mathcal{M} := \operatorname{supp}(\mu_\lambda) \neq \mathbb{R}^d$; then there exists a disc $B(x,\delta)$ whose intersection with $\mathcal{M}$ is empty. Hence by definition,

$$f_{i_1}^{-1} \ldots f_{i_n}^{-1} B(x,\delta) \cap \mathcal{M} = \emptyset$$

for any sequence of maps $(i_1,\ldots,i_n)$, $i_j \in \{1,2\}$. We have

$$f_{i_1}^{-1} \ldots f_{i_n}^{-1} B(x,\delta) = B(y_n, \lambda^n \delta),$$

where $y_n = f_{i_1}^{-1} \ldots f_{i_n}^{-1}(x) = \lambda^n M_{i_1}^{-1} \ldots M_{i_n}^{-1}(x)$. Since γ_1, γ_2 generate S^d, given any $\varepsilon > 0$ for arbitarily large n there exists a sequence $i_1,\ldots,i_n$ such that

$$\| M_{i_1}^{-1} \ldots M_{i_n}^{-1}(x) - \|x\|\bar{e}\, \| < \varepsilon. \tag{3.11}$$

Fix $r > 1$, $\varepsilon = \delta/2$ and n sufficiently large to satisfy $\lambda^n \geq 2r/\delta$ together with (3.11). Let $z = \lambda^n \|x\| \bar{e}$; we claim that

$$B(z,r) \subset B(y_n, \lambda^n \delta).$$

To see this suppose that $y \in B(z,r)$, i.e., $\|y - z\| \leq r$. Then

$$\|y - y_n\| \leq \|y - z\| + \|z - y_n\|$$

$$\leq r + \lambda^n \varepsilon = r + \frac{1}{2}\lambda^n \delta < \lambda^n \delta.$$

Hence $B(z,r) \cap \mathcal{M} = \emptyset$, and $f_3^{-k} B(z,r) \cap \mathcal{M} = \emptyset$ for any $k \geq 0$. Since $\|\bar{e}\| = 1$ and z belongs to the half-line $\{t\bar{e}, t \geq 0\}$, there exists $k \geq 0$ such that $f_3^{-k} B(z,r) \supset B(0, r-1)$. Hence for any $r > 1$, $B(0, r-1) \cap \mathcal{M} = \emptyset$, which means $\mathcal{M} = \emptyset$. This establishes the proposition. $\qquad \square$

References

[1] Alexander J. C. and Zagier D. (1991). *The entropy of a certain infinitely convolved Bernoulli measure*, J. London Math. Soc. **44**, 121–134.

[2] Barnsley M. F. and Elton J. H. (1988). *A new class of Markov processes for image encoding*, Adv. Appl. Prob., **20**, 14–32.

[3] Diaconis P. and Freedman D. (1999). *Iterated random functions*, SIAM Review **41**, 45–76.

[4] Dubins L. E. and Freedman D. A. (1966) *Invariant probabilities for certain Markov processes*, Ann. Math. Stat. **32**, 837–848.

[5] Erdős P. (1939). *On a family of symmetric Bernoulli convolutions*, Amer. J. Math. **61**, 974–975.

[6] M J Field. 'Generating sets for compact semisimple Lie groups', *Proc. Amer. Math. Soc.*, (2000)

[7] M Kuranishi. 'Two element generations on semisimple Lie groups', *Kodai. math. Sem. report*,(1949), 9–10.

[8] M. Nicol, N. Sidorov and D. Broomhead, 'On the fine structure of stationary measures in systems which contract-on-average', *to appear Journal of Theoretical Probability*

[9] Peres Y., Schlag W. and Solomyak B. (2000). *Sixty years of Bernoulli convolutions*, Fractals and Stochastics II, (C. Bandt, S. Graf and M. Zaehle, eds.), Progress in Probability **46**, 39–65, Birkhauser.

[10] Sidorov N. and Vershik A. (1998). *Ergodic properties of Erdös measure, the entropy of the goldenshift, and related problems*, Monatsh. Math. **126**, 215–261.

[11] Solomyak B. (1995). *On the random series $\sum \pm \lambda^i$ (an Erdös problem)*, Annals of Math. **142**, 611–625.

Matthew Nicol
Mathematics Department
University of Surrey
Guildford, Surrey GU2 5XH, United Kingdom
e-mail: M.Nicol@surrey.ac.uk

Nikita Sidorov
Department of Mathematics
UMIST
P.O. Box 88
Manchester M60 1QD, United Kingdom
e-mail: Nikita.A.Sidorov@umist.ac.uk

David Broomhead
Department of Mathematics
UMIST
P.O. Box 88
Manchester M60 1QD, United Kingdom
e-mail: David.Broomhead@umist.ac.uk

Trends in Mathematics:
Bifurcations, Symmetry and Patterns, 189–195
© 2003 Birkhäuser Verlag Basel/Switzerland

Rayleigh-Bénard Convection with Experimental Boundary Conditions

Joana Prat, Isabel Mercader, and Edgar Knobloch

Abstract. The onset of convection in systems that are heated via current dissipation in the lower boundary or that lose heat from the top boundary via Newton's law of cooling is formulated as a bifurcation problem. The Rayleigh number as usually defined is shown to be inappropriate as a bifurcation parameter since the temperature across the layer depends on the amplitude of convection and hence changes as convection evolves at fixed external parameter values. Moreover, the final state of the system is also different since it depends on the details of the applied boundary conditions. A modified Rayleigh number is introduced that does remain constant even when the system is evolving, and solutions obtained with the standard formulation are compared with those obtained via the new one.

1. Introduction

Rayleigh-Bénard convection has been the subject of a large number of theoretical and experimental studies since the pioneering work of Lord Rayleigh and H. Bénard. Yet despite this there are still various issues whose significance has not been fully recognized. One such issue that is important for the interpretation of experiments relates to the proper boundary conditions to be used in any theoretical treatment of the problem. In papers in which comparison between experiment and theory is attempted this important question is usually dismissed with the glib statement that boundaries consisting of a 'good' thermal conductor correspond to fixed temperature boundary conditions while those made of a thermal insulator correspond to fixed heat flux. However, rarely does the author consider the question of how good a conductor does the boundary have to be before the fixed temperature boundary conditions apply. In fact it appears that a commonly held view is that copper is a good thermal conductor and hence that copper boundaries are always correctly modeled by constant temperature boundary conditions. However, this is not so. Whether or not a boundary behaves like a constant temperature boundary does not depend only on its composition but also on its heat capacity relative to the heat capacity of the fluid with which it is in contact. The basic issue is fundamentally whether a change in the temperature of the fluid in contact with the boundary produces a small change in the temperature of the boundary or a large one. If it is the former then the boundary is effectively a constant temperature one; otherwise it is not. Both the conductivity of the material of the boundary (and of the confined fluid) and its mass enter into the determination of the (dimensionless) Biot number that characterizes the thermal properties of the boundary when in contact with a given mass of fluid. Regrettably, experimentalists almost never provide sufficient information to allow a theorist to estimate the Biot numbers for the lower and upper boundaries used in their experiment. It is this fact that is responsible for one of the main difficulties in comparing experiment with theory.

These issues are particularly acute when one studies problems in which the bifurcation to convection is *subcritical*. In this case once the conduction state loses stability the system evolves far from the initial state and the temperature difference ΔT across the layer drops, since convection decreases the temperature of the lower boundary and increases that of the upper boundary. In a bifurcation diagram showing the Nusselt number (a dimensionless measure of the heat transport) against the conventionally defined Rayleigh number (proportional to ΔT) the system therefore follows a path that slopes to the left, instead of evolving vertically as one would expect of a system under constant conditions. This behavior is shown quite dramatically in a number of the early experiments on binary fluid convection in which the lower boundary was heated electrically (via constant electrical power) [1, 2, 3].

In this note we indicate how the convection problem can be reformulated in order to define a bifurcation parameter that remains constant under fixed external conditions. A proper treatment of the problem requires the solution of a time-dependent conduction problem in boundaries of finite width, whose outer boundary may be at fixed temperature (if the boundary is held in thermal contact with a heat bath at fixed temperature) [4, 5]. Note that the whole notion of a heat bath requires that the thermal capacity of the bath be large compared to that of the sample. The issue here is that under typical experimental conditions the boundaries themselves do not act like a fixed temperature bath and instead conduct heat away from or into the fluid. A full discussion of this problem will be provided elsewhere [6]; here we follow [7] and focus on the derivation of the appropriate bifurcation parameter, assuming that the Biot numbers of the lower and upper boundaries are known.

2. The equations

We consider two-dimensional Boussinesq thermal convection between boundaries responsible for the boundary conditions

$$\frac{dT_-}{dz} = -\frac{B_-}{d}(T_L - T_-) \qquad \text{at} \quad z = -d/2,$$

$$\frac{dT_+}{dz} = -\frac{B_+}{d}(T_+ - T_U) \qquad \text{at} \quad z = d/2,$$

where d is the layer depth, T_L, T_U are the temperatures of the lower (L) and upper (U) heat baths, and T_-, T_+ are the temperatures in the fluid right next to the lower and upper boundaries (see fig. 1). Here $B_\mp$ are the Biot numbers of the boundaries: conducting boundaries correspond to $B = \infty$ while an insulating boundary corresponds to $B = 0$. Note that, by hypothesis, T_L, T_U are constants independent of time, while T_-, T_+ fluctuate in response to the motion of the fluid.

FIGURE 1. Sketch of the fluid layer

In the standard description of the Rayleigh-Bénard problem one describes the system in terms of the dimensionless temperature difference across the fluid in the conduction state, regardless of whether this state is stable or not. In this state $dT/dz = -\Delta T^c/d$ everywhere, and so

$$\Delta T^c = B_-(T_L - T_-^c) = B_+(T_+^c - T_U).$$

Here $T_\mp^c$ are the temperatures at the bottom and top of the fluid in the conduction state. Since $T_-^c - T_+^c = \Delta T^c$ it follows that

$$\Delta T^c = (T_L - T_U)\frac{B_+B_-}{B_+B_- + B_+ + B_-},$$

and is therefore independent of the dynamics of the system, provided $T_L - T_U$ remains fixed. Moreover, in view of the equivalent relation

$$\Delta T^c = \frac{B_+B_-\Delta T - d(B_+T_-' + B_-T_+')}{B_+B_- + B_+ + B_-},$$

the temperature difference $\Delta T \equiv T_- - T_+$ across the fluid may indeed change during evolution but must be compensated by a corresponding variation in the temperature gradients (indicated by a prime) at the top and bottom. We propose therefore to define a modified Rayleigh number based on the quantity ΔT^c (as opposed to ΔT) as a proper bifurcation parameter for the system.

To do this we nondimensionalize the equations in the usual way, expressing the temperature T in units of ΔT^c, distances in units of the layer depth d and time in units of d^2/κ, where κ is the thermal diffusivity of the fluid. We define the dimensionless control parameter by the relation

$$Ra'' = \frac{\alpha g(T_L - T_U)d^3}{\kappa\nu} = Ra'\frac{T_L - T_U}{\Delta T^c},$$

where Ra' is defined by analogy to the usual Rayleigh number, i.e., Ra' is proportional to the temperature difference across the fluid in the conduction state:

$$Ra' = \frac{\alpha g d^3}{\kappa \nu} \frac{B_+ B_- \Delta T - d(B_+ T'_- + B_- T'_+)}{B_+ B_- + B_+ + B_-}.$$

Note that both Ra' and Ra'' defined in this manner remain constant for fixed external conditions. Thus both provide good definitions of a control parameter, in contrast to the conventionally defined Rayleigh number

$$Ra = \frac{\alpha g \Delta T d^3}{\kappa \nu}.$$

Of the two Rayleigh numbers R' and Ra'' the latter may prove to be more useful since it is defined in terms of $T_L - T_U$ and not the temperature difference across the fluid.

In the following we list the final dimensionless equations describing convection in two dimensions, expressing the temperature fluctuation θ away from the conduction profile $T_-^c - \Delta T^c(z/d + 1/2)$ in units of ΔT^c. These are written in terms of a mean flow $\mathbf{U} = (U, 0)$ and its fluctuating part $\mathbf{v}' = (-\partial_z \chi', \partial_x \chi')$, where $\overline{\mathbf{v}'} = \overline{\chi'} = 0$, with the overline indicating an average over the horizontal period [8]. The result is

$$(\partial_t - \sigma \partial_{zz}^2)U + \partial_z \overline{v'_x v'_z} = 0, \qquad (1a)$$

$$(\partial_t + U \partial_x - \sigma \nabla^2)\omega' + Ra' \sigma \partial_x \theta + \partial_{zz}^2 U \partial_x \chi' + \frac{\partial(\chi', \omega')}{\partial(x, z)} - \overline{\frac{\partial(\chi', \omega')}{\partial(x, z)}} = 0, \qquad (1b)$$

$$(\partial_t + U \partial_x - \nabla^2)\theta - \partial_x \chi' + \frac{\partial(\chi', \theta)}{\partial(x, z)} = 0, \qquad (1c)$$

where $\omega' = -\nabla^2 \chi'$, and σ is the Prandtl number. Note that $Ra = Ra'(1 + \overline{\theta}_- - \overline{\theta}_+)$. The boundary conditions on the temperature fluctuation θ are

$$(1 - B_\pm^*)\partial_z \theta = \mp B_\pm^* \theta \quad \text{at} \quad z = \pm 1/2, \qquad (1d)$$

where $B_\pm^* = B_\pm/(1 + B_\pm)$. The modified Biot numbers $B_\pm^*$ are convenient for numerical exploration, and are such that $B = 0(\infty)$ corresponds to $B^* = 0(1)$. Thus $B^* = 1$ corresponds to a fixed temperature boundary condition, while $B^* = 0$ corresponds to a fixed flux boundary condition. For no-slip boundaries the velocity boundary conditions are

$$U = \chi' = \partial_z \chi' = 0 \quad \text{at} \quad z = \pm 1/2. \qquad (1e)$$

When $0 < B_\pm^* < 1$ the solution of the above problem depends on the choice of $B_\pm^*$ and hence differs from the corresponding solution with fixed temperature boundary conditions.

Fig. 2 shows the results obtained by solving equations (1a-e) for the case $(B_-^*, B_+^*) = (1, 0.8)$ and $\sigma = 0.1$ using a spectral Galerkin-Fourier technique in x and collocation-Chebyshev in z [9] in a periodic box of period $L = \pi$. The figure shows $N - 1$, where N is the Nusselt number, for the $n = 2$ state (solid curves) as a function of (a) Ra'' and (b) Ra, and compares the results with those for

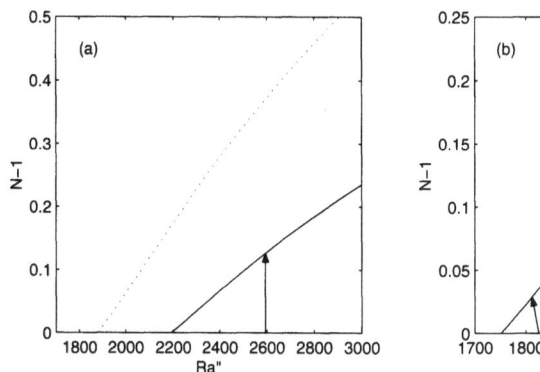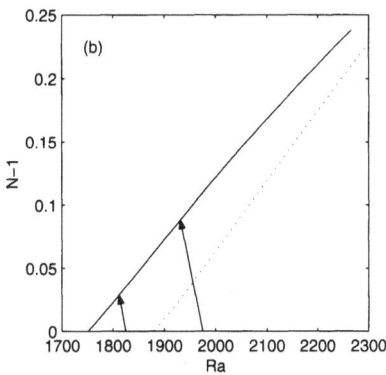

FIGURE 2. Bifurcation diagram for $L = \pi$, $\sigma = 0.1$ for the system (1a-e) with $B_-^* = 1$, $B_+^* = 0.8$ as a function of (a) Ra'' and (b) Ra. The primary instability is to an $n = 2$ mode (at $Ra'' = 2187$), followed by an instability to an $n = 1$ mode at $Ra'' = 2268$ (not shown). The dotted line shows the corresponding result for $B_+^* = 1$, and is identical in both plots. The arrows connect initial states with final states for several different initial conditions, and show that the conventionally defined Rayleigh number Ra decreases during evolution (fig.b). This is not the case in (a).

$B_-^* = B_+^* = 1$ (dotted curves). The evolution from the unstable conduction state that results is indicated by arrows. These are vertical in fig. 2(a) but slope to the left as in fig. 2(b) whenever $B_+^* < 1$. This slope can be used as a diagnostic for the value of the Biot number. In both cases the Nusselt number in the final state is independent of z and hence equals the Nusselt number N_+ at the top. For a perfect conductor at the bottom

$$N_+ - 1 = \frac{B_+^*}{1 - B_+^*} \frac{Ra' - Ra}{Ra'}.$$

Thus provided $B_+^* > 0$ we may use either $N_+ - 1$ or $(Ra' - Ra)/Ra'$ as indicators of the amplitude of convection. We emphasize that the solid and dotted curves are *not* the same, i.e., the assertion that Biot numbers only affect the evolutionary path in the Nusselt number-Rayleigh number diagram but not the final equilibrated state is manifestly false.

When $0 \leq B_-^* = B_+^* < 1$ the system retains midplane reflection symmetry but in the $N - 1$ vs Ra diagram it still evolves towards smaller Ra. All primary bifurcations necessarily produce branches of equilibria that are symmetric with respect to the two operations $T\kappa$ and R, where T denotes translation through half a wavelength, κ reflection in the midplane, and R a left-right reflection about a node. In contrast, the secondary (pitchfork) bifurcation on the $n = 1$ branch produces an unstable state with the (smaller) point symmetry $RT\kappa$, hereafter a

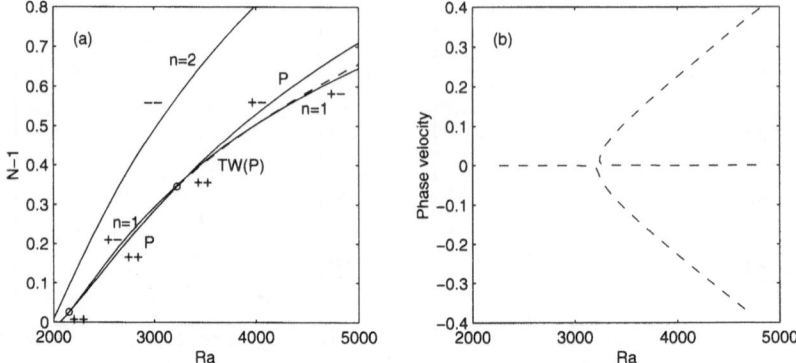

FIGURE 3. Bifurcation diagrams for (a) $B_-^* = B_+^* = 1$, and (b) $B_-^* = 1$, $B_+^* = 0.9999$, when $L = 2\pi/2.12$ and $\sigma = 0.1$. Solid (dashed) lines indicate steady (traveling) states. The loss of midplane reflection symmetry destroys the point symmetry of the P state and turns it into a slowly drifting state whose phase velocity is shown in (b). At the same time the tertiary parity-breaking bifurcation to TW seen in (a) turns into an imperfect bifurcation (see fig. b).

P state [10]. This state in turn undergoes a tertiary (pitchfork) bifurcation to a travelling wave state TW. Fig. 3(a) shows a diagram of this type for $B_-^* = B_+^* = 1$. When the Biot number at the top is changed slightly to break the symmetry κ the $n = 1$ and $n = 2$ states are essentially unchanged, but the P state turns into a slowly drifting traveling wave [10]; at the same time the pitchfork from the P state to the TW becomes imperfect, as illustrated in fig. 3(b).

With increasing asymmetry with respect to the midplane a range of Ra' develops in which none of the simple simple steady states or traveling waves just described is stable [11]. In this regime stable solutions with complex dynamics may be found [6].

3. Discussion

In this note we have pointed out a simple way of dealing with convection systems in which the effects of departures from perfectly thermally conducting boundaries affect the behavior of the system, and illustrated quantitatively these effects on both the linear stability analysis and the moderately nonlinear states of the system. The resulting formulation should be of substantial help for quantitative comparisons between experiments and theory.

Acknowledgements

This work was supported in part by DGESIC under grant PB97-0683 (IM and JP) and by the National Science Foundation under grant DMS-9703684 (EK). We thank the Fulbright Foundation for additional support.

References

[1] C.M. Surko, P. Kolodner, A. Passner and R.W. Walden, *Finite-amplitude traveling-wave convection in binary fluid mixtures*, Physica D, **23** (1986), 220–229.

[2] T.S. Sullivan and G. Ahlers, *Hopf bifurcation to convection near the codimension-two point in a ^3He-^4He mixture*, Phys. Rev. Lett., **61** (1988), 78–81.

[3] V. Steinberg, J. Fineberg, E. Moses and I. Rehberg, *Pattern selection and transition to turbulence in propagating waves*, Physica D, **37** (1989), 359–383.

[4] M.R.E. Proctor, *Planform selection by finite-amplitude thermal convection between poorly conducting slabs*, J. Fluid Mech., **113** (1981), 469–485.

[5] A. Recktenwald and M. Lücke, *Thermoconvection in magnetized ferrofluids: the influence of boundaries with finite heat conductivity*, J. Magn. Magn. Mater., **188** (1998), 326–332.

[6] I. Mercader, J. Prat and E. Knobloch, *Robust heteroclinic cycles in two-dimensional Rayleigh-Bénard convection without Boussinesq symmetry*. Preprint (2001).

[7] T.L. Clune, Ph.D Thesis, University of California at Berkeley (1993).

[8] J. Prat, J.M. Massaguer and I. Mercader, *Large-scale flows and resonances in 2-D thermal convection*, Phys. Fluids, **7** (1995), 121–134.

[9] J. Prat, I. Mercader and E. Knobloch, *Resonant mode interaction in Rayleigh-Bénard convection*, Phys. Rev. E, **58** (1998), 3145–3156.

[10] I. Mercader, J. Prat and E. Knobloch, *The 1:2 mode interaction in Rayleigh-Bénard convection with weakly broken midplane symmetry*, Int. J. Bif. Chaos, **11** (2001), 27–41.

[11] J. Prat, I. Mercader and E. Knobloch, *The 1:2 mode interaction in Rayleigh-Bénard convection with and without Boussinesq symmetry*, Int. J. Bif. Chaos, in press.

Joana Prat
Dept. Matemàtica Aplicada i Telemàtica
Universitat Politècnica de Cataluny, Barcelona, Spain
e-mail: joana@fa.upc.es

Isabel Mercader
Dept. Física Aplicada
Universitat Politècnica de Catalunya, Barcelona, Spain
e-mail: isabel@fa.upc.es

Edgar Knobloch
Department of Physics, University of California
Berkeley, CA 94720, USA

Trends in Mathematics:
Bifurcations, Symmetry and Patterns, 197–202
© 2003 Birkhäuser Verlag Basel/Switzerland

Global Bifurcations in FitzHugh-Nagumo Model

Adelina Georgescu, Carmen Rocşoreanu, and Nicolaie Giurgiţeanu

Abstract. The FitzHugh-Nagumo (F-N) system [1] modelling the electrical potential in the nodal system of the heart is shown to have a rich dynamics. The results are synthesized in the global bifurcation diagram providing an overall view of all possible qualitatively distinct responses of the model for all values of the parameters. Since the curves of global bifurcation values emerge at points of curves consisting of local bifurcation values, the global bifurcations are presented in the context of the global bifurcation diagram. Thus, codimension one bifurcations of Hopf, homoclinic, saddle-node, breaking saddle connections, nonhyperbolic limit cycle and breaking the connection between a saddle and a saddle-node types are obtained. A large number of codimension two bifurcations are discussed here, such as Bogdanov-Takens, Bautin, double homoclinic, double breaking saddle connections bifurcations. Some of the bifurcation boundaries are obtained analytically, other are obtained numerically, using the software MATHEMATICA and our own code DIECBI [2].

1. The F-N model

Among several models labeled F-N in the following the Cauchy problem $x(0) = x_0$, $y(0) = y_0$, for the second order ODE [1]

$$\dot{x} = c\left(x + y - x^3/3\right),$$
$$\dot{y} = -\left(x - a + by\right)/c. \tag{1}$$

will be refered to as the F-N model. The state functions $x, y : \mathbf{R} \to \mathbf{R}$, $x = x(t)$, $y = y(t)$ represent the electrical potential of the cell membrane and the excitability respectively, t is the time, $a, b \in \mathbf{R}$ are parameters depending on the number of channells of the cell membrane which are open for the ions of K^+ and Ca^{++} and $c > 0$ is the relaxation parameter.

The two–dimensional continuous time dynamics generated by this problem strongly depends on parameters. On the other hand, the complexity of the phase portraits is due to the presence of the equilibria, their invariant manifolds and orbits connecting them and the limit cycles. The main features of the corresponding dynamics is revealed by the global bifurcation diagram corresponding to the parametric portrait (Section 2). Our study concerns the case $c = const$, $c > 1$. In fact, in [3] we considered the restriction $c \geq 1 + \sqrt{3}$, in order to separate in the (b, a)-plane not only the regions with nontopological equivalent dynamics, but also the regions where the attractors (or repulsors) are foci or nodes, cases that are interesting for the biologists, but which are topological equivalent from a mathematical point of view. The case $c \leq 1$ complicates in addition due to the fact that the curve of Hopf bifurcation values changes drastically with c. Such a detailed discussion was carried out later and can be found in [5].

2. Bifurcations in F-N system

In the following we quote the local as well as global bifurcations for the F-N system obtained by the authors [3].

The saddle-node bifurcation takes place when parameters (b, a) are situated on the curves $S_{1,2}$ of equations

$$a = \pm \frac{2}{3} |b| \sqrt{\left(1 - \frac{1}{b}\right)^3}, \quad b \in (-\infty, 0) \cup [1, \infty), \tag{2}$$

except at the points

$$Q_1 = \left(-c, \frac{2}{3}(c+1)\sqrt{1 + \frac{1}{c}}\right), \quad Q_3 = \left(c, \frac{2}{3}(c-1)\sqrt{1 - \frac{1}{c}}\right), \quad Q_2 \text{ and } Q_4$$

their symmetrics with respect to the Ob-axis respectively, and $Q = (1, 0)$.

For parameters situated on these curves, the system possesses a double equilibrium point and the system linearized around it has an eigenvalue equal to zero. At the point Q, the single equilibrium point is the origin, which is a non-hyperbolic repulsor. At the points Q_i, $i = 1, 2, 3, 4$, the system possesses a double equilibrium with a double zero eigenvalue and a Bogdanov-Takens bifurcation takes place.

The Hopf bifurcation takes place for parameters (b, a) situated on the curves $H_{1,2}$ of equations

$$a = \pm \frac{b}{3}\left(-2 + \frac{3}{b} - \frac{b}{c^2}\right)\sqrt{1 - \frac{b}{c^2}}, \quad b \in (-c, c). \tag{3}$$

At two exceptional points

$$Q_{17,18} = \left(c^2 - c\sqrt{c^2 - 1}, \pm \frac{4}{3}\left(c\sqrt{c^2 - 1} - c^2 + 1\right)\sqrt[4]{1 - \frac{1}{c^2}}\right) \in H_{1,2} \tag{4}$$

the Hopf bifurcation degenerates into a Bautin bifurcation.

Let $Q_0 = H_1 \cap H_2$. At Q_0 two Hopf bifurcations take place simultaneously around two equilibria.

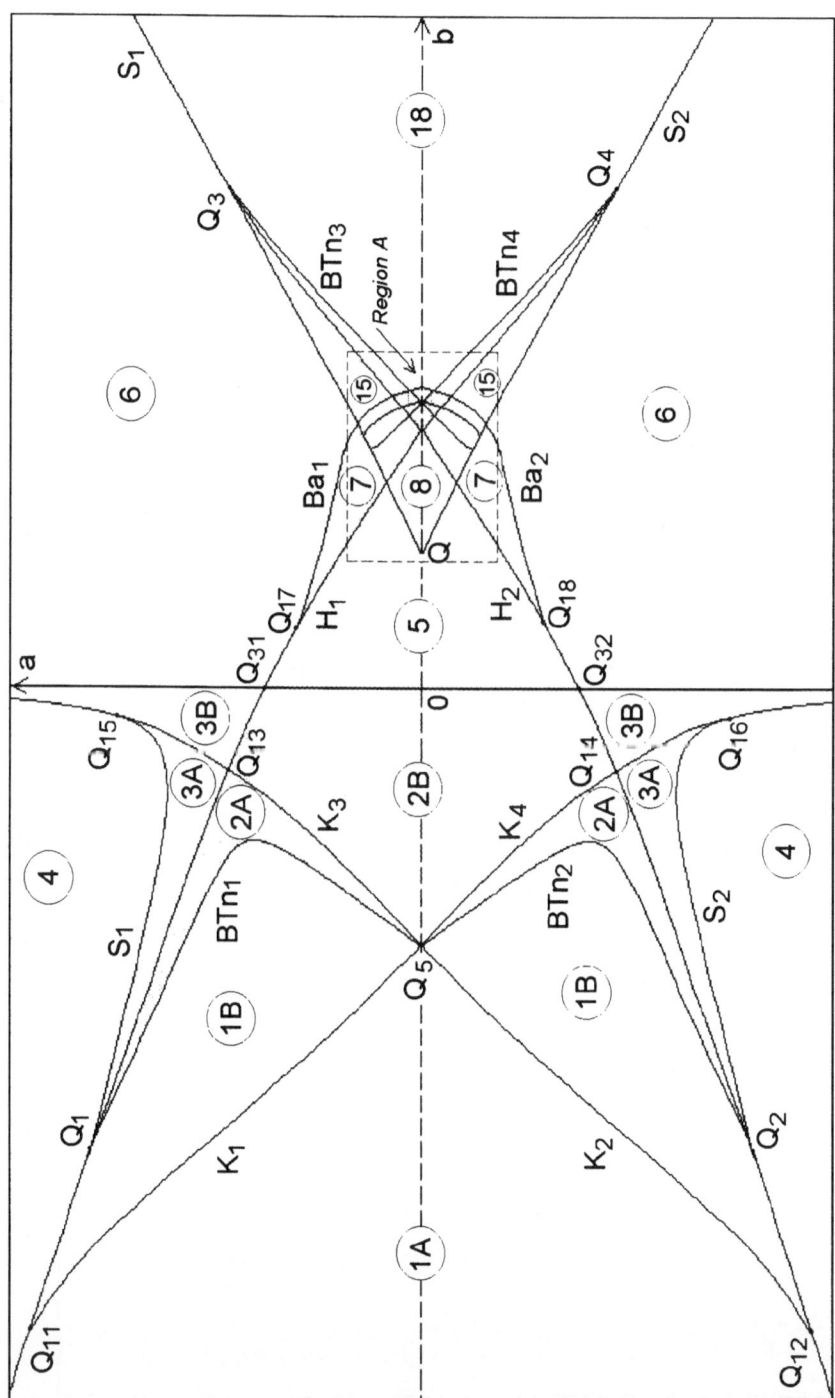

Fig. 1. Parametric portrait of the FitzHugh-Nagumo system. Shaded regions contain limit cycles.

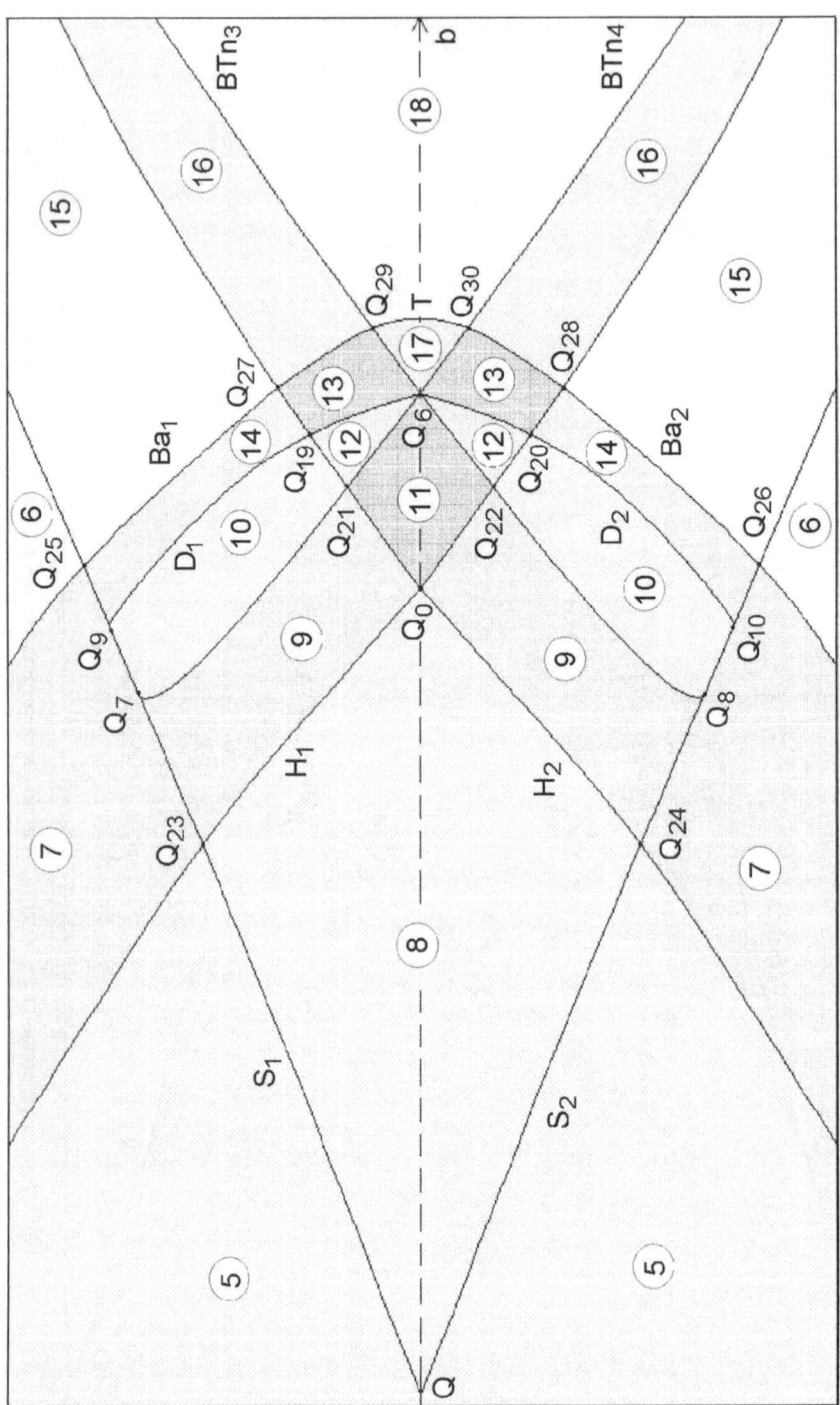

Fig. 2. Region A from Figure 1. Shaded regions contain limit cycles, darker shading for larger number of cycles.

Using the normal form method [4], we found that the equation of the curves $Ba_{1,2}$, emerging at $Q_{17,18}$, has the following asymptotic expression in the neighborhood of these points

$$\left[\operatorname{Re} c_2 - \frac{1}{\omega} \operatorname{Re} c_1 \operatorname{Im} c_1 + \mu \left(\frac{1}{\omega^2} \left(\operatorname{Im} c_1 \right)^2 - \frac{1}{\omega} \operatorname{Im} c_2 \right) \right] \tag{5}$$

$$\times \left(\frac{1}{\omega} \operatorname{Re} c_1 - \frac{\nu}{\omega} \operatorname{Im} c_1 \right)^2 + 4\mu = 0,$$

where c_1, c_2 are related to the Liapunov coefficients, μ and ω are the real and imaginary parts of the eigenvalues of the system linearized around the Bautin equilibrium point and $\nu = \frac{\mu}{\omega}$. For parameters situated on these curves a non-hyperbolic limit cycle bifurcation, (i.e., emerged as two hyperbolic limit cycles coalesce) takes place.

Also using the normal form method [4], we showed [3] that at the points Q_i, emerge the curves BTn_i, $i = 1, \ldots, 4$ corresponding to homoclinic bifurcations.

Their equations are approximated by

$$a = \pm \frac{7b^2 - 10bc^2 + 3c^2}{15c^3} \sqrt{\frac{7b^2 + 5c^2 b - 12c^2}{5b}}, \quad b \in (-c, 0). \tag{6}$$

$$a = \pm \frac{7b^2 + 10b\, c^2 - 17c^2}{15c^3} \sqrt{\frac{-7b^2 + 5b\, c^2 + 2c^2}{5b}}, \quad b \in (0, c). \tag{7}$$

Near the Bautin and Bogdanov-Takens bifurcation values respectively the curves Ba_i and BTn_i were drawn by asymptotic formulae up to points at which they were coincident to the results of the direct numerical computations. From these points on, these curves were drawn by the computations. In this way we derived the domain of validity of asymptotic formulae.

The curves BTn_3 and BTn_4 intersect at Q_6, where a double homoclinic bifurcation takes place. The curves BTn_1 and BTn_2 interesect at Q_5 where a double breaking saddle connection bifurcation (i.e., two saddle connections are broken) takes place. In addition at Q_6 emerge the curves $D_{1,2}$, obtained only numerically [2], and consisting of homoclinic bifurcation values. At Q_5 emerge the curves K_{1-4} obtained also only numerically and consisting of breaking saddle connection bifurcation values. The curves $D_{1,2}$ and $BTn_{3,4}$ cut the curves $S_{1,2}$ at Q_{7-10}, where a saddle-node separatrix loop bifurcation takes place, whilst the curves K_{1-4} cut the curves $S_{1,2}$ at $Q_{11,12}$ and $Q_{15,16}$, where saddle–node–saddle with separatrix connection bifurcations take place [6]. Among all these bifurcations, the global ones correspond to the parameters situated on the curves BTn_{1-4}, $D_{1,2}$, K_{1-4}, $Ba_{1,2}$.

Putting together the local bifurcation diagrams obtained for the quoted bifurcations, the global parametric portrait is obtained [3] (Fig. 1, 2). The points Q_{13}, Q_{14} and Q_i, $i = 19, \ldots, 30$ in Figures 1,2 are other codimension two bifurcation points. Regions with the same label have identical or symmetrical phase portraits.

In the activity of the heart the periodic beatings are of crucial importance. They correspond to limit cycles occurring for parameters situated in the (b, a)-plane in the shaded regions from the parametric portrait. Thus, regions 2A, 2B, 5, 8, 10, 16 correspond to a single limit cycle, regions 7, 9, 12, 14, 17 correspond to two limit cycles, whilst regions 11 and 13 correspond to three coexisting limit cycles. The completion of the parametric portrait with schematic phase portraits and various types of oscillations corresponding to different shapes of the limit cycles are given in [3]

References

[1] FitzHugh, R. (1961) Impulses and physiological states in theoretical models of nerve membrane, Biophysical J, 1, 445–446.

[2] Giurgiteanu, N. (1997) Computational economical and biological dynamics – DIECBI, "Europa" Publisher, Craiova, (Romanian).

[3] Rocsoreanu, C., Georgescu, A., Giurgiteanu, N. (2000) FitzHugh-Nagumo model. Bifurcation and dynamics, Kluwer Academic Publishers, Dordrecht.

[4] Kuznetzov, Yu. (1995) Elements of applied bifurcation theory, Springer, New-York.

[5] Sterpu, M., Georgescu, A. (2001) Codimension three bifurcation for the FitzHugh-Nagumo system, Mathematical Reports, Romanian Academy, 3 (53) 3.

[6] Rocsoreanu, C., Giurgiteanu, N., Georgescu, A. (2001) Connections between saddles for the FitzHugh-Nagumo system. Int. J. Bif. and Chaos, Vol 11, 2, 533–540.

Adelina Georgescu
Dept. of Math., University of Pitesti
00300 Pitesti, Romania

Carmen Rocşoreanu
Dept. of Math., University of Craiova
A.I. Cuza, 13
1100 Craiova, Romania

Nicolaie Giurgiţeanu
Dept. of Ec. Sci., University of Craiova
A.I. Cuza, 13
1100 Craiova, Romania

List of Contributors

A. Alonso (Catalunya, Spain), M. Net, J. Sánchez

P. Ashwin (Exeter, UK)

A. Bayliss, **B. Matkowsky** (Northwestern, USA), A. Aldushin

T. Callahan (Michigan, USA)

G. Cicogna (Pisa, Italy)

S. Cox (Nottingham, UK), **P. Matthews** (Nottingham, UK)

M. Field (Houston, USA)

S. van Gils (Twente, The Netherlands), O. Diekmann

P. Glendinning (UMIST, UK)

M. Golubitsky (Houston, USA) and **D. Chillingworth** (Southampton, UK)

A. Goetz (San Francisco, USA), **M. Mendes** (Surrey, UK)

P. Matthews (Nottingham, UK), **S. Cox** (Nottingham, UK)

M. Nicol (Surrey, UK), N. Sidorov, D. Broomhead

J. Prat (Catalunya, Spain), I. Mercader (Catalunya, Spain),
E. Knobloch (Berkeley, USA)

C. Rocsoreanu, **A. Georgescu** (Pitesti, Romania), N. Giurgiteanu

A. Rucklidge (Cambridge, UK), **M. Silber** (Northwestern, USA), J. Fineberg

D. Rusu (Guelph, Canada), **W. Langford** (Guelph, Canada)

I. Stewart (Warwick, UK), T. Elmhirst, J. Cohen

List of Participants

Stella **Abreu**
Departamento de Matemática
Universidade Portucalense
R.Dr. António Bernardino
de Almeida, 541/619
4200 Porto, Portugal
scabreu@fc.up.pt

Manuela **Aguiar**
Faculdade de Economia
Universidade do Porto
Rua Dr. Roberto Frias
4200 Porto, Portugal
maguiar@fep.up.pt

Arantxa **Alonso**
Department of Applied Physics
Politechnic University of Catalonia
Jordi Girona Salgado, s/n
Campus Nord, Modul B-5
08034 Barcelone, Spain
arantxa@fa.upc.es

Peter **Ashwin**
School of Math Sciences
University of Exeter
Exeter EX4 4QE, UK
PAshwin@maths.ex.ac.uk

Dwight **Barkley**
Mathematics Institute
University of Warwick
Coventry CV4 7AL, UK
barkley@maths.warwick.ac.uk

Jorge **Buescu**
Departamento de Matemática
Instituto Superior Técnico
Av. Rovisco Pais
1046-001 Lisboa, Portugal
jbuescu@math.ist.utl.pt

Pietro-Luciano **Buono**
Mathematics Institute
University of Warwick
Coventry CV4 7AL, UK
buono@maths.warwick.ac.uk

Timothy **Callahan**
University of Michigan
2072 East Hall
525 East University
Ann Arbor, MI 48109-1109, USA
timcall@math.lsa.umich.edu

Sofia **Castro**
Centro de Matemática Aplicada
Faculdade de Economia
Universidade do Porto
Rua Dr. Roberto Frias
4200 Porto, Portugal
sdcastro@fep.up.pt

David **Chillingworth**
Department of Mathematics
University of Southampton
Southampton SO17 1BJ, UK
drjc@maths.soton.ac.uk

Vladimir **Chiricalov**
Department of Mech. and Math.
Nat. T.Shevchenko Univ.
Volodymyrs'ka str. 64
04033, Kyiv, Ukraine
chvo@mechmat.univ.kiev.ua

Pascal **Chossat**
INLN
1361 route des Lucioles
Sophia Antipolis
06560 Valbonne, France
chossat@inln.cnrs.fr

Giampaolo **Cicogna**
Dipartimento di Fisica
Universitá di Pisa
Via Buonarroti 2, Ed. B
56127 – Pisa, Italy
cicogna@df.unipi.it

Gabriela **Constantinescu**
Constantin Bratescu College
Constanta
Str. Rascoalei
8700 Constanta, Romania
gabiconstantinescu@usa.ne

Eurico **Covas**
Astronomy Unit
School of Mathematical Sciences
Queen Mary and Westfield College
Mile End Road
London E1 4NS, UK
E.O.Covas@qmw.ac.uk

Stephen **Cox**
School of Mathematical Sciences
University of Nottingham
University Park
Nottingham
NG7 2RD, UK
stephen.cox@nottingham.ac.uk

Jonathan **Dawes**
Department of Applied Mathematics
and Theoretical Physics
University of Cambridge
Silver Street
Cambridge, CB3 9EW, UK
J.H.P.Dawes@damtp.cam.ac.uk

Ana Paula **Dias**
Departamento de Matemática Pura
Centro de Matemática
da Universidade do Porto
Praça Gomes Teixeira
Faculdade de Ciências
Universidade do Porto
4099-002 Porto, Portugal
apdias@fc.up.pt

Benoit **Dionne**
Department of Mathematics
and Statistics
University of Ottawa
585 King Edward Avenue
Ottawa
Ontario K1N 6N5, Canada
benoit@mathstat.uottawa.ca

Bernold **Fiedler**
Freie Universität Berlin
Fachbereich Mathematik
und Informatik
Arnimallee 2-6
D-14195 Berlin, Germany
fiedler@math.fu-berlin.de

Michael **Field**
Department of Mathematics
University of Houston
Houston, TX 77204-3476, USA
mf@uh.edu

José Orlando Gomes **Freitas**
Departamento de Matemática
Universidade da Madeira
Largo do Colégio
9000 Funchal, Portugal
orlando@uma.pt

Adelina **Georgescu**
Dept of Applied Mathematics
University of Pitesti
Str. Tirgul din Vale, 1
0300 Pitesti, Romania
cristi@geostar.ro

Stephan **van Gils**
Faculty of Mathematical Sciences
University of Twente
P.O. Box 217
7500 AE Enschede, The Netherlands
stephan@math.utwente.nl

Paul **Glendinning**
Department of Mathematics
UMIST
P.O. Box 88
Manchester M60 1QD, UK
p.a.glendinning@umist.ac.uk

Arek **Goetz**
Department of Mathematics
San Francisco State University
1600 Holloway Ave
San Francisco, Ca 94080, USA
goetz@sfsu.edu

Martin **Golubitsky**
Department of Mathematics
University of Houston
Houston, TX 77204-3476, USA
mg@uh.edu

Gabriela **Gomes**
Ecology and Epidemiology Group
Department of Biological Sciences
University of Warwick
Coventry CV4 7AL, UK
M.G.M.Gomes@warwick.ac.uk

Edgar **Knobloch**
Department of Physics
University of California
Berkeley, CA 94720-7300, USA
knobloch@physics.berkeley.edu

Martin **Krupa**
Department of Mathematical Sciences
New Mexico State University
Las Cruces, NM 88003-8001, USA
mkrupa@nmsu.edu

Isabel **Labouriau**
Centro de Matemática Aplicada
Departamento de Matemática Aplicada
Rua das Taipas, 135
4050-600 Porto, Portugal
islabour@fc.up.pt

Jeroen **Lamb**
Department of Mathematics
Imperial College
London, SW7 2BZ, UK
jsw.lamb@ic.ac.uk

William **Langford**
Mathematics
and Statistics Department
University of Guelph
Guelph Ontario Canada
N1G 2W1, Canada
wlangfor@msnet.mathstat.uoguelph.ca

Míriam **Manoel**
Departamento de Matemática
ICMC-USP
Caixa Postal 668
13560-970 São Carlos, SP, Brasil
miriam@icmc.sc.usp.br

Bernard **Matkowsky**
Dept. of Eng. Sci.
and Appl. Math.
Northwestern University
2145 Sheridan Rd.
Evanston, IL 60208, USA
b-matkowsky@nwu.edu

Paul **Matthews**
School of Mathematical Sciences
University of Nottingham
University Park
Nottingham NG7 2RD, UK
paul.matthews@nottingham.ac.uk

Ian **Melbourne**
Department of Mathematics
and Statistics
University of Surrey
Guildford, GU2 5XH, UK
I.Melbourne@surrey.ac.uk

Miguel **Mendes**
Department of Mathematics
and Statistics
University of Surrey
Guildford
Surrey GU2 5XH, UK
mmendes@fe.up.pt

Wayne **Nagata**
Department of Mathematics
University of British Columbia
121 – 1984 Mathematics Road
Vancouver, B.C., Canada V6T 1Z2
nagata@math.ubc.ca

Matthew **Nicol**
Department of Mathematics
and Statistics
University of Surrey
Guildford, GU2 5XH, UK
m.nicol@surrey.ac.uk

Jesus **Palacian**
Departamento de Matemáticas
Universidad Pública de Navarra, Spain
palacian@unavarra.es

Eliana **Pinho**
Departamento de Matemática-Informática
Universidade da Beira Interior
Rua Marquês d'Ávila e Bolama
6200-001 Covilhã, Portugal
eliana@noe.ubi.pt

Alberto Adrego **Pinto**
Centro de Matemática Aplicada
Departamento de Matemática Aplicada
Rua das Taipas, 135
4050-600 Porto, Portugal
aapinto@fc.up.pt

Carla **Pinto**
Rua 9 Julho 52 1o B
4050-433 Porto, Portugal
carlampinto@mail.telepac.pt

Joana **Prat**
Despatx 113
Departament de Matemática Aplicada
i Telemática
Universitat Poli'ecnica de catalunya
Avda. Victor Balaguer s/n
E.U.P.V.G, 08800 Vilanova
i la Geltru, Spain
joana@mat.upc.es

Jens **Rademacher**
Freie Universität Berlin
Fachbereich Mathematik
und Informatik
Arnimallee 2-6
14195 Berlin, Germany
jens.r@earthling.net

Mark **Roberts**
Department of Mathematics
and Statistics
University of Surrey
Guildford, GU2 5XH, UK
M.Roberts@surrey.ac.uk

Carlos **Rocha**
Departamento de Matemática
Instituto Superior Técnico
Av. Rovisco Pais
1046-001 Lisboa, Portugal
crocha@math.ist.utl.pt

Hildebrando Munhoz **Rodrigues**
SMA – ICMSC – USP
Caixa Postal 668
São Carlos – SP 13560-970, Brasil
hmr@icmc.sc.usp.br

Alastair **Rucklidge**
Department of Applied Mathematics
and Theoretical Physics
University of Cambridge
Silver Street
Cambridge CB3 9EW, UK
A.M.Rucklidge@damtp.cam.ac.uk

Jennifer **Siggers**
Department of Applied Mathematics
and Theoretical Physics
Silver Street
Cambridge CB3 9EW, UK
jhs21@cam.ac.uk

Mary **Silber**
Engineering Sciences
and Applied Mathematics
Northwestern University
2145 Sheridan Road
Evanston, IL 60208, USA
m-silber@nwu.edu

Esmeralda **Sousa Dias**
Departamento de Matemática
Instituto Superior Técnico
Av. Rovisco Pais
1049-001 Lisboa, Portugal
edias@math.ist.utl.pt

Ian **Stewart**
Mathematics Institute
University of Warwick
Coventry CV4 7AL, UK
ins@maths.warwick.ac.uk

Rob **Sturman**
Centre for Nonlinear Dynamics
and its Applications
Department of Civil Engineering
University College London
Gower Street
London WC1E 6BT, UK
ucesrjs@ucl.ac.uk

Harry **Swinney**
Physics Department and
Center for Nonlinear Dynamics
University of Texas at Austin
Austin, Texas 78712, USA
swinney@chaos.ph.utexas.edu

Claudia **Wulff**
Freie Universietät Berlin
Fachbereich Mathematik
und Informatik
Arnimallee 2–6
14 195 Berlin, Germany
wulff@math.fu-berlin.de

Patricia **Yanguas**
Departamento de Matemática
e Informática
Universidad Pública de Navarra
Campus de Arrosadia
31006 Pamplona (Navarra), Spain
yanguas@unavarra.es

In Preparation:
New softcover edition
ISBN 3-7643-2171-7

Golubitsky, M., University of Houston, USA /
Stewart, I., University of Warwick, UK

The Symmetry Perspective
From Equilibrium to Chaos in Phase Space and Physical Space

2002. 342 pages. Hardcover
ISBN 3-7643-6609-5
PM - Progress in Mathematics, Vol. 200

Winner of the Sunyer prize 2001

Pattern formation in physical systems is one of the major research frontiers of mathematics. A central theme of *The Symmetry Perspective* is that many instances of pattern formation can be understood within a single framework: symmetry.

The symmetries of a system of nonlinear ordinary or partial differential equations can be used to analyze, predict, and understand general mechanisms of pattern-formation. The symmetries of a system imply a "catalogue" of typical forms of behavior, from which the actual behavior is "selected".

The book applies symmetry methods to increasingly complex kinds of dynamic behavior: equilibria, period-doubling, time-periodic states, homoclinic and heteroclinic orbits, and chaos. Examples are drawn from both ODEs and PDEs. In each case the type of dynamical behavior being studied is motivated through applications, drawn from a wide variety of scientific disciplines ranging from theoretical physics to evolutionary biology. An extensive bibliography is provided.

Buescu, J., Instituto Superior Técnico, Lisaboa, Portugal

Exotic Attractors
From Liapunov Stability to Riddled Basins

1997. 144 pages. Hardcover
ISBN 3-7643-5793-2
PM - Progress in Mathematics, Vol. 153

This book on attractors in dynamical systems will appeal primarily to researchers and advanced postgraduate students working in the area of dynamical systems. However, since it is self-contained, it may be used profitably by anyone wishing a general but mathematically rigorous introduction to the concepts and ideas of attractors in dynamics.

The study is divided roughly into two parts, with a generic introduction to the concept of attractor in dynamics preceding a description of new results on two research problems. The first part is gentle but rigorous; several different notions of attractor are defined and compared, and the finer points are thoroughly illustrated by examples and counterexamples.

The second part of the book deals with two different problems in discrete dynamics to which the author has contributed. One is the characterization of the dynamics on stable ω-limit sets with infinitely many components; this is shown to be an adding machine, which has interesting implications for dynamics at a fundamental level. The second problem is the study of the transverse stability of attractors on an invariant submanifold.

ANALYSIS WITH BIRKHÄUSER

For orders originating from all over the world except USA and Canada:

Birkhäuser Verlag AG
c/o Springer GmbH & Co
Haberstrasse 7
D-69126 Heidelberg
Fax: ++49 / 6221 / 345 42 29
e-mail: birkhauser@springer.de

For orders originating in the USA and Canada:

Birkhäuser
333 Meadowland Parkway
USA-Secaucus
NJ 07094-2491
Fax: ++1 201 348 4505
e-mail: orders@birkhauser.com

http://www.birkhauser.ch